D1337074

DISCOVERY

IN RUSSIAN AND SIBERIAN

WATERS

DISCOVERY

IN RUSSIAN AND SIBERIAN

WATERS

L. H. NEATBY

OHIO UNIVERSITY PRESS : ATHENS

A NOTE TO THE READER

Few spheres of peaceful human activity excite more general interest than geographical discovery, and this is particularly true of exploration in the harsh and forbidding zones which encircle the Poles. Two centuries of abundant literature telling of discovery in the American Arctic and Antarctica testify to this truth. In comparison pioneer work in the eastern or Russian Arctic has been neglected. There are a number of reasons for this: things Russian have always seemed alien and remote to the Western world. The history of northeastern pioneering lacks the unity bestowed on exploration in the American Arctic by the underlying motive of the Northwest Passage and in Antarctica by the rapid sequence of such masterful adventurers as Scott, Shackleton, Amundsen and Byrd. The Russians were poor publicists: the Tsarist regime viewed its officers with suspicion and mistrust. Its treatment of Bering and Wrangell is in sharp contrast to the honors conferred on Cook in the eighteenth century and on Peary and Shackleton in our own. Nor was there an informed public and press in Russia. And, until the present century, discovery in the Russian Arctic was largely due to the uncoordinated efforts of men of various nations, who represented no national purpose or aspiration and who aroused no sustained national interest. Thus, the explorers of the North Pacific, of the Siberian shore, and of the ocean above it did not command the applause which made popular heroes of several generations of pioneers in other parts of the icy zones.

Yet the heroism was there, and this book is a modest attempt to bring it to light. No one could be more acutely conscious of its insufficiency than the author himself: his limited resources, advancing years and professional duties in a wholly alien field must serve as his apology. These pages are a labor of love, and one can only hope that they will furnish the reader with some of the enjoyment which the writer had in compiling them.

The author's thanks are due to the library staffs of Acadia University, Wolfville, Nova Scotia, the Department of Northern Development, Ottawa, and the University of Saskatchewan, Saskatoon, especially for their help in xeroxing and obtaining books. He is grateful to Professors W. O. Kupsch, R. Williamson, and W. Barr all of the Institute of Northern Studies, University of Saskatchewan, for their help and encouragement, and particularly to Mrs. Phyllis Smith, secretary to the Institute, for supervising and sharing in the typing of the script. He also records his indebtedness to Mr. Wayne Macmillan of the College of Engineering for preparing the maps, and to the Medical Services for reproducing illustrations.

Although this book is largely based on the explorers' original records, the author owes much to the industrious research of F. A. Golder, especially for his edition of Steller's journal of his voyages with Bering, to Robert Murphy's lively treatment of this theme in *The Haunted Journey*, and to Swen Waxell's *The American Expedition*, for one or two details of Bering's last voyages unnoticed by Steller. Special acknowledgment is due to the Soviet explorer and historian V. Y. Viese and to his book *Morya Sovetskoe Arktiki*, a comprehensive survey of Russian Arctic history which generously recognizes the work done by all foreigners in the Russian Arctic. Thanks are also due to Dr. J. C. Beaglehole for his edition of *The Journals of Captain James Cook*, Vol. 3, with its enlightening introduction, and to Ann Parry for her study *Parry of the Arctic*.

Institute of Northern Studies
University of Saskatchewan

vi

CONTENTS

DISCOVERY

IN RUSSIAN AND SIBERIAN

WATERS

THE ENGLISH: EARLY EXPLORERS IN
THE RUSSIAN ARCTIC

The frozen seas to the north of Europe and Asia rank next to the sources of the Nile as the region which has for the longest period engaged, without satisfying, the curiosity of civilized man. For our first reliable notice of this area we are indebted to a Greek of Marseilles named Pytheas. This man, notable as being perhaps the first to compute latitude by the sun's shadow, was, in about 325 B.C., sent by his native city to report on the commercial possibilities of northern Europe; besides making a survey of the coast of Gaul, he visited (and named) the British Isles and gathered what information he could of even more remote countries. From the people of Britain he learned of the land of Thule which lay six days' voyage to the North on the fringe of everlasting fields of ice. Unfortunately the account written by this enterprising Greek has been lost; he is known to us only through his fellow countrymen, Polybius and Strabo, who disbelieved his report and treated him with the contempt which scientists of all ages reserve for the pioneer who has travelled too far beyond the frontier of ascertained fact. But time has restored his credit, and his Thule, long identified with Iceland, is supposed by some to be the remoter parts of Norway.

The journey of Pytheas belongs to the later period of ancient Greece, when the vitality of that extraordinary race was dying away. The Roman, who succeeded the Greek as the standard-

3

bearer of Western culture, had less inclination to pursue knowledge which promised neither profit nor glory. New geographical facts gathered by the Romans through commercial transactions were not definite, and probably were not systematically recorded. The next positive attempt to push back the northern frontier belongs to the ensuing Dark Ages. As a part of his campaign to mitigate the ignorance and barbarism of his age Alfred the Great of England translated into Saxon the general history of the late Roman, Orosius, to which he made sundry additions, including discoveries from the recent travels of his Norwegian-born vassal Ohthere (Ottar).

Ohthere, so he told the king, "dwelt the furthest north of any Norman," (in Halgoland) with no settlement beyond his abode except those of a few scattered Finns, who lived the primitive life of hunters and fishermen. The Northman "fell into a fantasie and desire to prove and to know how farre that land stretched Northward, and whether there were any habitations of men North beyond the desert"—perhaps Ohthere was the first man to undertake a long journey purely for scientific reasons. Three days after setting sail he reached the limit of the whale fisheries; in another three he found that the coast fell away to the east and then "bowed directly towards the South" into the White Sea. For five days he coasted into the gulf, observing on his starboard a desert land sparsely settled by a few hunters and fishermen. Coming to the mouth of a mighty river and a well peopled land, he began to feel the uneasiness natural to a man of limited experience. The folk to whom he had come—the Biarmes—were more civilized than the Finns and were tolerably good agriculturists; nonetheless Ohthere would not trust himself in their power. He thought they spoke the same language as the Finns; evidently he did not have sufficient mastery of that tongue to be positive. Ohthere also gave the king a rough idea of the topography of the upper Scandinavian peninsula and of the whale fisheries, which even then (A.D. 888–890) seem to have been a thriving industry.

4

But Ohthere, as far as our knowledge extends, remained as unique and exceptional a figure as the king who was his patron. Others, no doubt, followed and extended the course which Ohthere had embarked upon: Nansen supposes it reasonable that during the Middle Ages a trade grew up between the Norwegians and the Russians who dwelt on the southern and eastern shores of the White Sea, but those who conducted it found no Alfred to put their journeys on record. Extension of firm knowledge of the north of Russia had to await the coming of the Modern Age.

In the sixteenth century the marvelous energies released by the Renaissance were in full play, promoting intellectual awakening and artistic creation in the Old World, discovery, exploitation and settlement in the New. Up to that time the material prizes had been ill-distributed—by virtue of discovery, occupation and Papal decree—Spaniards and Portuguese divided control of the Americas and the Indies and had set up a barrier against the traders of northern Europe, which was only partially breached half a century later by force. The French challenged this monopoly by setting up a short-lived colony in Florida and by harrying Spanish shipping in the West Indies; the English, hindered from such methods by the still-existing alliance with Spain, were forced to seek a commercial outlet in new and less promising regions. In the first instance this was found in the Russian Arctic.

With the approval, therefore, of the government of Edward VI, presided over by the regent Northumberland, the merchants and gentlemen of London formed an association for discovery and trade in the little-known and forbidding regions which lay to the northeast beyond the north cape of Norway. They entitled themselves "the Companie of the Marchants adventurers for the discoverie of Regions, Dominions, Islands and places unknowen"; and their ultimate purpose was to find a short route to China through the Russian Arctic. This object was never achieved; but useful purposes were effected nearer home, and posterity remembers the association by the brief and accurate designation of the "Muscovy Company."

This enterprise found wide backing among the well-to-do: shares of twenty-five pounds each were sold to raise a total fund of 6,000 pounds; three ships were specially constructed with a view to the peculiar conditions of the intended cruise; and the command of the expedition was entrusted to a soldier who had gained distinction in the Scottish wars, Sir Hugh Willoughby. The Elizabethans saw nothing odd in promoting to such an appointment as sea captain a landsman who had little or no knowledge of nautical matters. The tradition of the feudalism which the Tudors had overthrown was still strong: the ordinary seaman would pay readier obedience to a gentleman than to one of his own sort who had no other superiority than technical skill. In any case the authority of the Captain General was severely restricted. He was merely the president and chief executive of a board of twelve Counsellors, on which the nautical profession, the mercantile interest and the Church were all represented, and by which every important decision was to be made and approved. Perhaps a diversified command was suited to so experimental a voyage, but a generation later that great innovator, Sir Francis Drake, helped to affirm the sounder principle of an absolute and undivided command by chopping off the head of a troublesome and possibly treacherous counsellor.

Willoughby's squadron consisted of three ships: himself, the Captain General, in the *Bona Esperanza* (120 tons), with one Jefferson as master (the master was a skilled professional seaman); the *Edward Bonaventure* (160 tons), Richard Chancellor, Captain, accompanied by Stephen Borough as master, and the *Confidentia* (90 tons), commanded by Cornelius Durfurth. The orders for the expedition, drawn up by the aged sailor-geographer, Sebastian Cabot, were to become a model for all the nations of Europe. One recognizes the old seaman in the elaborate and lengthy instructions for the preservation of harmony, discipline and good conduct on board, the directions that competent persons on each ship make daily observations of landmarks, tides, soundings, the altitude of the sun and the position of moon and stars, and, whenever possible, should compare observations to arrive at a full and trustworthy record for the guidance of future adventurers. It is worth noting that though these orders were entitled "ordinances, instructions and advertisements of and for the intended voyage for Cathay," Cabot makes no direct reference to China, while giving abundant advice as to the proper bearing toward more primitive cultures who might be encountered on the way. The old traveler knew better than his partners how hard China was to arrive at and also perceived that the best market for English broadcloth lay not in the Celestial Empire but in the sub-Arctic parts of North Russia which lay between England and it. The English adventurers had little of the religious fervor which distinguished the Spanish in the New World: their emissaries were commanded "not to disclose to any nation the state of our religion, but to pass it over in silence, without any declaration of it, seeming to beare with such lawes and rites, as the place hath, when you shall arrive." However sharp the religious strife at home, the Englishman would not permit it to hamper his commercial operations abroad.

These orders were dated the 9 May 1553. King Edward VI furnished Willoughby with letters of commendation to the "Kings, Princes, and other Potentates, inhabiting the Northeast partes of the Worlde, towardes the mighty Empire of Cathay."

And, all preparations having been made, the ships cast off from Deptford on May 10 and, dropping down stream, on the following day they passed Greenwich Palace where the Court then resided. (Twenty-three years later Queen Elizabeth was there to wave farewell to Martin Frobisher who was embarking for the North *West* Passage.)

Once at sea the expedition was not slow in experiencing the setbacks which the sober-minded had foreseen. The primitive three-masters in which they were embarked were not good at sailing into the wind; and northerly gales detained them for many days on the English east coast. They remained long at Harwich, where leaking wine casks and putrefying food added to the cares of the leaders. Delay and inaction produced their natural consequence in a decline of morale; seamen expressed increasing reluctance to prosecute a journey which so quickly had lost its glamour. Chancellor himself was troubled by the thought of his two little sons, "Which were in the case of Orphanes if he speede not well"; he found it difficult to inspire his men to encounter dangers which he himself had begun to dread.

The expedition had quitted the Thames in mid-May. Not until July 14 did it make its landfall on the coast of Norway, a little below the Arctic Circle. From the twenty-seventh to the thirtieth they were kindly entertained by the natives of the Lofoten Isles, who, however, could furnish no directions for the course to be steered. The end of August found them at Seynam Island in 70° north latitude. The natives undertook to furnish a pilot to the Danish fortress in Finnmark, the Wardhouse, (Vardo), a promise which, if redeemed, might have saved the lives of the whole company. But while the ships lay to in the offing they were smitten by a tempest which, "the land being very high on every side," scattered them by "flawes in the winde and terrible whirlwindes," and forced them to seek the open sea. The Admiral's pinnace, launched to communicate with the island, foundered; Willoughby hailed Chancellor with an earnest request not to go far from him; but the *Edward* was a sluggish sailing vessel, and the crew of the *Esperanza*, unable to control their ship properly, soon dropped their consort from sight. For Willoughby's crew it was a dreadful night of fog and tempest, steering blind amid a maze of rock and shoal. At daybreak the fog lifted and, perceiving a ship to leeward, they made a light sail and drew down to her. She proved to be the *Confidentia*. They never found the *Edward* again.

Willoughby's subsequent course has proved a puzzle to those who have attempted to reconstruct it. It had been agreed earlier that in the event of separation the ships were to rendezvous at Wardhouse, but the fleet's best pilots, Chancellor and Borough, were both on the *Edward*; and no one remained with the Captain General capable of taking him to a port. It appears that, failing to arrive at the rendezvous and realizing that he was too far to the east, Willoughby resolved to leave his chief pilot behind and continue the voyage to China. On August 14 he sighted land in 72° north latitude (the coast of Novaya Zemlya) and hoisted out a boat to investigate but shoal water and shore ice made landing impossible. With his easterly course blocked and the *Confidentia* leaking badly, the Admiral ran before a northeasterly wind back to the Lapland shore, and there spent much time plying up and down that unfriendly coast seeking a safe harbor and friendly settlements. Twice his ships were driven out to sea and twice they recovered the land. On September 18 they found a harbor both sheltered and deep. The wondering sailors saw "seale fishes, and other great fishes, and upon the maine we saw beares, great deere, foxes, with divers strange beestes, as guloines, which were to us unknown and also wonderfull." The shore was barren and uninhabited. Frost, snow and hail assailed them as if in the depth of winter, yet the tired wayfarers had not the resolution to quit the secure anchorage and again challenge the unfriendly sea. Parties were sent out in three directions to seek natives who might give help and advice for the hard winter months which lay ahead. One after another they straggled back "without finding of people, or any similitude of habitation."

Willoughby's journal ends with this despairing sigh. Want of skill and inconstancy of purpose had led him into difficulties; want of adaptability made the difficulties fatal. It was the opinion of an English merchant familiar with the region that, had Willoughby, instead of wintering on board ship, built snow houses in a sheltered spot on shore, he and his men might have lived through the winter; as it was they "perished for cold." It appeared from a will found among Willoughby's papers that he and most of his party were still alive in January, 1554. "One must needs say they were men worthy of a better fortune," wrote Chancellor's chronicler in a phrase that is forceful by its very restraint.

Richard Chancellor, in the meantime, on finding himself separated from the Admiral, steered directly for Wardhouse and there in conformity with his orders waited seven days. Europeans had already found their way to that Finnish outpost, among them certain Scots who warned the Englishman in the friendliest manner to pursue his journey no farther, "amplifying the dangers which hee was to fall into." Chancellor, however, was not disposed to give up simply because he had a nearer view of dangers which he had foreseen from the first; and his men, emboldened by the perils they had surmounted, bade him go forward. So forward they went, over a sea where with the delighted surprise of children they "found no night at all, but a continual light and brightness of the sunne shining clearly upon the huge and mightie Sea." Willoughby had been led away by the hope of a quick voyage to Cathay; his chief pilot harbored no such illusions. He felt his way down the Lapland coast, turned in at the mouth of the White Sea and found an anchorage at the west mouth of the river Dvina, near the modern Archangel. The natives, impressed by the size of his ship, did reverence to him as a god, but he "took them up in all loving sort from the ground," gaining thereby for himself and his company a character for condescension and gentleness. He learned that he was in Russia or Moscovie, and that his new friends were the subjects of the Tsar Ivan Vasilivich, "the Terrible." *

* The *Edward Bonaventure* cast anchor in the Dvina estuary at a time when the Russian people had emerged from the chaos of the Tartar invasion and set foot on the pathway to empire. From ancient times the Slavic branch of the Indo-European family had been settled on the steppes to the north of the Black Sea and astride the great trade route which conducted from the Baltic to the Black Sea by way of Lake Ladoga up the Volkhov and down the Dnieper. The civilizing influences of Greek and Iranian culture which trade helped to introduce were offset by a geographical handicap which would have extinguished a less sturdy race: the Slavic community had no natural frontier; its flat and treeless terrain made it the corridor through which time and again the hordes of Asia passed to conquest in the west. The Slavs, however, held out against this scourge by patient endurance, stubborn toil and a birthrate that was proof against

The English: Early Explorers in the Russian Arctic

Chancellor soon discovered that the Tsarist authority was a reality, even on the Arctic shore. The local governor received him with courtesy and undertook to provide him and his crew with food but would by no means permit the stranger to sell his wares

all discouragement, their abundant population enabling them to flow in large numbers into any vacuum left in the wake of the barbarian conquerors. In the years following the fall of the western half of the Roman Empire, Slavs reached the North Sea at Hamburg. They were thrown back and assimilated into Roman Catholic Christianity as the principalities of Poland and Bohemia. The South Slavs poured into the Balkan Peninsula, and were cut off from the parent tribe by the wedge driven into Europe by the Hungarian Magyars. In this isolation they survived with true Slavic toughness centuries of Turkish oppression to vex the politics of modern Europe. The East Slavs, the Russians, who remained by their ancestral waterways, were organized into a coherent state in the ninth and tenth centuries by the Norse Vikings, Rurik and Helgi. Orderly government produced its fruit in vigorous trade. Novgorod in the north, Kiev in the centre, and Kherson on the Black Sea, were links in a chain of commerce extending on to Constantinople (Istanbul), where the successors of the Roman Caesars still held sway. Russia was Christianized, trading posts on the Volga linked her with the older civilization of Persia, and it seemed that the strong and vigorous rule of the Norseman would do more for the Slav in the East than it had done for the Anglo-Saxon in the West.

The hopeful promise of a Russian civilization, linked by way of Constantinople with the Catholic culture of the West, was wrecked by the catastrophe of the Tartar invasion, beginning in A.D. 1222. In the north Novgorod was shielded by its marshes, but the southern part of the water route and the capital city of Kiev were engulfed. The vanquished Russians fell back on the familiar resources of passive endurance and the healing operations of time. Moscow, a town on the fringe of the forest zone, removed from the water route, but conveniently situated for the control of both it and the Volga channel, became the nucleus of a new Russian state. Here the prince of the house of Rurik set up his court and to it migrated the Metropolitan of the Russian Church and the better sort of displaced persons from Kiev. The prince of Moscow sought to extend his authority as the chief vassal of the conquering power, and the Tartars somewhat short-sightedly countenanced this policy that he might serve as their agent in collecting tribute from the subject Russians. The latter, for their part, submitted readily enough to a prince on whose influence they relied to protect them against fresh barbarian inroads. The ruler of Moscow gained rapidly in territory and prestige. In 1453 the capture of Constantinople and the destruction of the Byzantine Empire by the Turks made him the first layman in the Greek Church; he consolidated this claim by wedding Sophia, kinswoman of the last Byzantine emperor, Constantine Palaelogus, and by appropriating his title "Tsar" (Caesar). In 1475 he was strong enough to seize forcibly the city and principality of Novgorod, and in 1480 he declared and successfully maintained his independence of the Tartar Horde. The new Russia had arrived.

But though extensive and populous the new monarchy was worse equipped than the old principality of Kiev for keeping pace with the march of Western culture. During the long crisis of Tartar oppression, Church and laity had of force conceded their prince powers more despotic and also more permanent than those which, in a not dissimilar situation, the English people had granted to the monarchs of the Tudor line. The decrees of the Tsars were executed throughout their dominions by a body of officials, who, if sometimes corrupt, were seldom inefficient and were a far more servile or undertake the journey to Moscow without leave obtained from the Tsar. Ivan was not slow to grant this permission. Apart from his dependence on the goodwill of Poland, he was suffering from the greed of the Baltic towns of the Hanseatic League, who controlled Russian trade to the West and were abusing their monopoly to the point of extortion. Chancellor's voyage gave promise of cheaper goods and freer communication, so Ivan sent a cordial message to the English captain granting permission to trade, and inviting him to visit Moscow at the expense of the Imperial government. In the meantime by the threat of pulling his ship out and returning to England, Chancellor had prevailed on the governor to consent to his immediate departure for Moscow. He was met on the way by the emissaries of the Tsar and conducted with every mark of honor to the Muscovite capital.

Twelve days elapsed before he was admitted to the royal presence; he spent the interval in touring Moscow, observing, inquiring and commenting on all that he learned with eager curiosity and unprejudiced objectivity. After detailing the products of different regions, the centers of industry and commerce, roads and the means of transport, he describes life in the capital at some length, and it is interesting to note that, though to our way of thinking life in Tudor England was harsh, primitive and subject to oppression, Chancellor is as confident of its superiority to the Russian way of life as any modern tourist from Britain or America might be, though like the present-day traveler he notes the tremendous military potential of the vast and populous Rus-

---

body than the merchants and gentlemen from whom the Tudor civil servants were recruited. Furthermore Russia had to a great extent lost touch with the rest of Europe. Constantinople and the Black Sea outlet were now in the hands of the infidel Turk. With the ebbing of the Tartar flood Kiev and the southern part of the water road had fallen under the domination of the Poles whose power, extending from the Baltic to the Black Sea, walled them in with a barrier which for several generations no Tsar was able to break down. Frontal attack, however, was not the Russian method. The Tsar avoided an out-and-out struggle with the Poles, and was securing his rear by wringing control of the Volga and the Caspian Sea from the Tartars, when word was received by Ivan the Terrible that English traders, arriving by way of the White Sea, promised to restore the unfettered communication with Europe which the Tartar invasion had cut off.

sian state. While criticizing the Russian for his lack of order and discipline, he cannot contain his admiration for his capacity to endure months of campaigning with no other protection than the clothes he wears in weather which "our boasting warriors" would find unbearable. "Now what might be made of these men if they were trained and broken to order of civill wars (civilized methods of warfare) ?" . . . "If they knewe their strength no man were able to make match with them. . . . But I thinke it is not God's will." It did not occur to him that the ultimate effect of the traffic he was promoting might be to defeat the supposed will of the Almighty.

Most amusing is the way in which Chancellor, a subject of the most autocratic government ever in the British Isles, prides himself on the contrast between the liberties enjoyed by the humblest Englishman and the servitude of the Russian noble. The "Duke," he tells us, has the power to confiscate any man's goods and redistribute them among more favored vassals. All that the victim can say is that "he hath nothing, but it is Gods and the Dukes Graces, and cannot say, as *we the common people in England*, if we have anything, that it is Gods and our owne." At the same time he is impressed by the Russian's patriotism and passionate loyalty to his prince: he thinks it an honor to serve in the army at his own cost, instead, as is the case in England, of using his rank, money, or influence to evade military service. Chancellor also notes with apparent approval that "they have no Lawyers, but every man is his owne Advocate." The honest sailor had no jurist at his elbow to suggest that perhaps the liberties on which he prided himself owed their form and permanence to the legal profession.

Once admitted to the Imperial presence Chancellor readily obtained permission to do business in the Tsar's dominions; he sailed back to England the next summer with letters offering to the English company trading privileges, which were exactly stated and solemnly confirmed in an edict issued by the Tsar some months later. But the brave Chancellor was not to share in the profits of the commerce which his enterprise had created. In the late summer of 1556 he embarked on his second return voyage from the White Sea bearing with him the Russian ambassador, Osep Napea. Nearing the Scottish coast Chancellor's buffeted ship could not weather the easternmost point of Aberdeenshire; he anchored, but was beaten from his ground tackle by an "outrageous tempest" and driven among the rocks of Aberdour Bay. Still mindful of his country's honor Chancellor got the Russian envoy ashore, but his boat was swamped and he with many of his men died in the surf. One can only hope that the "two little sonnes" of whose orphanhood he had had an anxious premonition were well cared for by the Company of Merchants he had served so well.

The merchants of London did not allow the prospects of the Russian trade to divert them from the task of continued exploration toward the Far East. In 1556 Stephen Borough, who had been Chancellor's master on his first voyage to the White Sea, was assigned the task of further discovery. Rather than risk a large vessel and crew in the incalculably dangerous work of exploration the Company appointed him to the pinnace *Serchthrift*, an excellent choice, apart from her limited cruising radius, as Borough was time and again to experience. Her handiness and light draft gave her crew confidence in her on the coast of the Barents Sea whose shoals and havens, though well known to the native fishermen, had never been recorded on map or chart. Sebastian Cabot, a very old man now, but still alert and thorough, visited the *Serchthrift* at her dock and approved her equipment. In conformity with the pious and kindly custom of the period he distributed "liberall alms" to poor folk at the quayside, "wishing them to pray for the good fortune and prosperous successe of the voyage."

Borough sailed in company with trading vessels bound for the White Sea under command of Chancellor, now on his last outward voyage, around the North cape. There in the gloom of a dense fog he heard the signal gun on the Admiral's ship which warned him to quit the convoy and steer a course for the east. Borough, who doubtless learned a good deal about those regions while wintering on the Dvina in 1553–1554, first steered for the mouth of the river Kola (Kolui) where he anchored his ship to effect repairs. There, as he probably expected, he met thirty Russian lodias, twenty-oared river and fishing craft, now bound for the waters east of the river Pechora where they spent the summer fishing and hunting the morse (walrus). The skipper of one of these vessels, a man named Gabriel, struck up a great friendship with the English captain, gave him a barrel of mead and one of beer, "which was caryed upon mens backes at least two miles," and promised to guide him and warn him of shoals on the voyage to the Pechora.

On Monday, June 22, 1556, all set sail for the west side of the Kanin Peninsula. With a light breeze the flat-bottomed riverboat quickly outdistanced the sturdy oceangoing pinnace, and on the next day Borough found himself in difficulties in what he complains was not sea but "sunke land, and full of shoales," bottom at two fathoms, no land in sight, and threatening signs of a gale from the north. Gabriel, who with rare humanity had lagged behind the convoy to keep the stranger in sight, came to his relief and piloted the *Serchthrift* to a creek where the fishing boats had already found haven. As she was crossing the bar the wind failed and Borough was compelled to anchor at the harbor mouth to avoid drifting ashore. He borrowed two small anchors and with the zealous aid of Russian friends, "who were likely to have bene drowned for their labour," tried to warp his way to a secure anchorage within the harbour. But the ebbing tide threatened to leave him aground, so he slipped his cable, made for the deep water and rode out the gale in the open sea. As soon as he returned to harbor Gabriel came aboard to report that the Russians had already salvaged their own anchors and the *Serchthrift*'s tackle. With him came another captain, Keril, who had furnished one of the anchors, and wished to retain Borough's hawser in payment for that service. Desiring neither to submit to such extortion nor to provoke a quarrel, Borough handled the situation with tact: he showed Gabriel every courtesy while ignoring Keril, and when the latter had returned to his ship sent word after him that unless the hawser was returned immediately, he would complain to the Russian authorities. This resolute and courageous attitude won Keril's respect: he complied at once; put on his best coat and a collar of pearls, and came aboard with a gift, which was courteously received. He told Borough that "his father was a gentleman, and that he was able to shew me pleasure, and not Gabriel, who was but a priests sonne." They parted the best of friends.

Plying northward Borough worked his way down the west side of the Kanin Peninsula, availing himself of the ebb tide to make headway against the wind and anchoring in the flow to hold the distance he had gained. He envied the ease with which the Russian lodias found rest and shelter in creeks, while his men had no respite from labor and discomfort in the open sea. On the 9th of July he rounded Cape Kanin and on its eastern side found deeper water and stronger winds, generally from the north. He was in danger of a lee shore when Gabriel came again to his assistance and conducted him to the harbor of Morgoviets, where he obtained water, stones for ballast, and observed the variation of the compass. From there a short leg brought him to the region of low sandhills around the mouth of the Pechora. He parted from the hospitable Russians who had now reached their fishing grounds, and set a course east-northeast, still using the tide: "we stopped the ebbes, and plyed all the floods to the windewards." On the twenty-first he thought that he saw land; it proved to be "a monstrous heape of ice"; they were soon in the heart of the pack, "which was a fearful sight to see," and for six hours it was as much as they could do to avoid collision with one mass without crashing into another. "And when we had passed from the danger of this ice we lay to the Eastwards close to the wind." On the twenty-second the pinnace was nearly capsized by a whale which came so close aboard that they might have thrust a sword into him, "which we durst not doe for feare hee should have overthrown our shippe"; so Borough called together the crew, and "with the crie that we made he departed from us." That same day they sighted land and came to anchor on an island near the Kara Strait. Borough took observations and ascertained that the compass variation was much increased.

Ahead of him lay Vaygach Island and Kara Strait; to the north an unbroken mass of land. Misled doubtless by the sight of ice ahead into thinking that he was embayed and that there was no exit to the east he hauled to the wind and began to coast westward along the land seen in that direction, until he met a lodia whose captain informed him of his error and that the large land to the north was the separate formation of Novaya Zemlya. Borough again put about and still fighting contrary winds entered Kara Strait. In a heavy gale on the last day of July he found harborage on the north shore of Vaygach Island.

Two Russian lodias had taken refuge in the same bay, and from these Borough learned that the natives of that region, the Samoyeds, dwelt beyond the effective control of the Tsar. They were barbarous, living in tents of deerskin draped over poles and stakes. They supported themselves by hunting and rearing reindeer, did not practice agriculture, and had no bread except that obtained by trade. In the vicinity of Vaygach their manners had been somewhat softened by contact with Russian traders; further east, toward the river Ob, they treated strangers with murderous hostility. Borough went ashore with his new acquaintances who showed him a heap of some three hundred Samoyed idols, "the worst and most unartificiall (inartistic) worke that ever I saw".
. . . "Some of their idols were an olde sticke with two or three notches made with a knife in it"—the work of some abstract artist born before his time. One of the Russians spoke of going on the Ob if he had poor success in hunting walrus, but with great reluctance owing to the utter savagery of the eastern Samoyeds.

Northerly gales and much ice drove the *Serchthrift* again down the west shore of Vaygach Island where she lay embayed for several days in a tempest of snow and hail. The wind-beaten pack seaward was a "fearful sight to behold." Northeast winds blew continuously, bringing down more ice, and nights were lengthening with alarming rapidity. Either eastward progress was impossible or Borough's courage failed him. It was no light thing to venture into the icebound Kara Sea, off a shore infested with bloodthirsty savages and behind a strait liable at any time to be barred by ice. After some dodging from harbor to harbor in storm and mist Borough decided to return. He left Vaygach Island on August 22, and on September 11 anchored at Colmogro on the White Sea, and there spent the winter.

It had been Borough's intention to make another attempt at reaching the mouth of the Ob in the summer of 1557, but he received orders from Mr. Killingworth, the Company's agent, to go in search of two missing freighters. So he went up the Lapland coast, taking soundings, fixing and describing landmarks and anchorages with his accustomed diligence, until he arrived at Wardhouse. There he found representatives of a nation that was to be England's commercial rival in all quarters of the globe, and during the decline of Spain to contend with her in many bitter battles for the dominion of the seas. Dutchmen had chartered Norwegian vessels to establish a trade with Lapland. They received the Englishman with hospitality and courteously evaded his inquiries as to the rates at which they exchanged their wares for the furs and fish of Lapland. Borough pays his tribute to their commercial acuteness: "The Dutchmen bring hither mightie strong beere. I am certaine that our English double beere would not be liked of the Kerils and Lappians as long as that would last." From these men he learned that one of the ships he sought had been wrecked and the other, after wintering in Lapland, had sailed for England. And so his cruise terminated.

Traversing as he did waters that were frequented by Russian fishermen, Borough can hardly be called an explorer. But his unwearying industry in the accurate observation of soundings, tides, landmarks and harbors, and also his success in communicating with native populations make him a hydrographer of merit and a worthy forerunner of James Cook, the first scientific navigator to enter the Russian Arctic from the east, as Borough was from the west. His achievement did not go unnoticed. He received an invitation to Spain, then the foremost nation in the science and practice of navigation, and was courteously entertained by grandees and savants at Seville. In retrospect this visit was a melancholy affair, for religious differences and colonial rivalry were soon to break forever the friendship which in the Middle Ages had united the proud and adventurous peoples of England and Spain.

Little further progress was made by the English in the Russian Arctic. In 1580 Captain Arthur Pet in the forty-ton *George*, accompanied by one Jackman in the twenty-ton *William*, did force the passage of the Kara Strait to the north of Vaygach Island, but a few days' trial satisfied both captains that there was no way east through the Kara Sea that season. The persevering Pet worked his *George* among drifting ice blocks that mounted above the ship's maintop, while his phlegmatic colleague lay anchored to a floe, and "did as much good as we that did labour." They may have come back by way of the Yugorskiy Shar, thus circumnavigating Vaygach Island. An equally futile voyage by Henry Hudson to the west shore of Novaya Zemlya twenty-eight years later was the last attempt of the English to navigate the Northeast Passage.

THE DUTCH: BARENTS IN THE ARCTIC

On a dull and blustery afternoon in the autumn of 1871 the Norwegian sailing-vessel *Solid* of Hammerfest found itself off the northeast angle of Novaya Zemlya in 76° 12′ north latitude. As twilight came on her, Captain Elling Carlsen made out on shore a wooden structure, distinguished by its sharp rectangular lines from the smooth and rounded rocks on which it was built. Because of the contrary winds he landed to make a thorough investigation of the ruined shelter. It was a crude house built of driftwood and measuring sixteen meters by ten. He collected a quantity of relics, cooking pans of copper, a crowbar, a gun barrel, a clock, a flute, along with various articles of dress, and built a cairn to record his visit. Putting again to sea he made his exit with great difficulty by way of Kara Strait and arrived at Hammerfest, where the relics were purchased by an English tourist, Lister Kay. If Kay had any ungenerous impulse to make profit of his bargain, he quickly suppressed it. He turned over the articles for what they cost him to the Royal Dutch Museum, where they properly belonged. For the rude shack from which they had been gathered was the wintering place of the Dutch discoverers, Barents and Heemskirk, two hundred and seventy years before.

Barents's voyages mark the peak of Holland's endeavor in the Russian Arctic. Borough had met their traders near Murmansk, and they lost no time in challenging English monopoly of traffic by way of the White Sea. For a time, their commerce was almost paralyzed by the war of independence against Spain. In 1593, when the crisis of that struggle was past, the merchants of Middelburg prepared to send two ships to China by way of the Kara (Borough) Strait, while the Amsterdammers fitted out a third ship to make a similar attempt around the northern extremity of Novaya Zemlya. The command of the latter was entrusted to a most skillful pilot, Willem Barents. The two expeditions parted company near the North Cape. Barents crossed the sea which now bears his name and coasted northeast up the Novaya Zemlya shore, noting wreckage as evidence that Russians had penetrated thus far in pursuit of walrus. On July 29 they reached Ice Point but were barred by ice and gale from rounding it and proving that they had attained the northern extremity of the hyphenated peninsula of Novaya Zemlya. Near Kolguyev Island on the homeward track they met the Middelburg ships, whose commander, Cornelius Nai, boasted that they had penetrated two hundred miles into the Kara Sea and reached the longitude of the Ob River. Modern opinion endorses the judgment of Barents's friend and historian, Gerrit de Veer, that this was a "very broad" claim.

The States General of the United Provinces and the stadtholder, Prince Maurice, were more credulous. They financed a most ambitious expedition equipped for commerce as well as discovery. In 1595 six ships prepared to make the eastern voyage richly laden and accompanied by a seventh, a pinnace, which was to bring back word if and when the cargo vessels rounded Cape Tabin (the Taymyr Peninsula), which was vaguely but correctly supposed to be the northernmost point of Eurasia. The expedition found Kara Strait ice-choked and Vaygach Strait badly obstructed. Though they lingered for three weeks, their voyage, except for some restless maneuvering in the strait, had come to an end. Barents, the pure scientist, was urgent that they should make all possible progress through the broken fields of ice: his colleagues, concerned for valuable freight, were not inclined to move without a really promising opening. They perceived that the English had been right in separating the functions of explorer and merchant and in sending out Borough and Pet with the cheapest equipment consistent with efficiency. Westerly winds threatened to close the strait behind them, the weather grew "mistie, melancholy and snowie." The leaders had been waiting for some such sign: in the middle of September they put about for home, ingloriously, but without material loss.

In consequence of this failure the States General reluctantly declined any further attempt at public cost, but offered generous reward to any private association or individual who could prove the Eastern Passage viable. In 1596 the Amsterdammers fitted out two ships for a fresh trial and like good practical Hollanders manned them as far as possible with single men, "that they might not be disswaded by means of their wives and children to leave off the voyage"—married men were disinclined to the neck-or-nothing effort which, it had become evident, was needed to achieve success. Jacob Heemskirk was appointed to the command of the vessel on which Barents shipped as pilot, possibly to control his supposed rashness. The appointment was perhaps reasonable and certainly harmless: the two men cooperated with the same harmony as did, at a somewhat later date, the Englishmen, Bylot and Baffin. Their consort was commanded by John Cornelison Rijp.

They sailed May 10 and, setting a northerly course, were off the Shetlands by the twenty-second and by the first of June were deep within the Arctic Circle, in the zone of perpetual daylight. There was some debate between Rijp and Barents, Rijp desiring to set a more westerly course than suited the chief pilot. The former carried his point, and on June 8 discovered Bear Island, killed a bear swimming in the water, and passed a dead whale which "stouncke monsterously." The two ships pushed on, skirting the western edge of a vast icefield. In 80° north latitude they found land—Spitsbergen—which they at first supposed to be part of Greenland and sailed up its western side until stopped by ice. They noted on shore leaves, grass and "such plants as beasts feed upon," though on the Novaya Zemlya shore four degrees further to the south, scanty grass only was found and carnivorous bears and foxes.

It was then a common belief (which survived into the nineteenth century) that even in the highest latitudes the sea did not freeze at a distance from land. In application of this theory Rijp wished to push on further north. Barents entirely disagreed. So the ships parted company. After fruitless exertions to the east of Greenland Rijp returned home before winter set in.

In the meantime Heemskirk and Barents had steered the easterly course which the latter declared would bring them to Novaya Zemlya. On the seventeenth of July they reached Admiralty Island and began to coast up the Novaya Zemlya shore until on the nineteenth they were pinned against the land by an impenetrable ice field. A party on shore were approached by a bear with plainly aggressive intent. Heemskirk with great presence of mind and by threatening to stab the first person who offered to run held his men in a compact body and by shouting and uproar kept the bear at a distance and made a safe and orderly retreat to the boat. After battling for two weeks against wind and ice they reached Ice Point on August 6 and from there were driven back by contrary winds and on the fifteenth anchored the ship to ice near Orange Island. A number of men boated over to the Novaya Zemlya mainland and from the top of a hill ascertained that beyond Ice Point the coast trended southeast and then directly south. Ice Point was proved the northernmost extremity of Novaya Zemlya and away to the east stretched the Kara Sea.

"We were much comforted againe, thinking that wee had woon our voyage and knew not how we should get soon inough on boord to certifie William Barents thereof," wrote the eager diarist, Gerrit de Veer. But it was to prove a barren triumph. On rounding Ice Point and passing the Cape of Desire (Soviet cartographers have preserved this pathetic name in the Russian form of Cape Zhelaniya) Barents found the coastline remorselessly bending away from a south to a southwesterly trend toward Vaygach Island instead of pointing southeast to Siberia and China. The ice of the Kara Sea, too, proved heavy and impenetrable, making progress down the east shore of the Zemlya peninsula impossible. Northeasterly winds robbed them of the last resource of getting back around Ice Point: on August 26 wind and ice forced them to seek a permanent harbor in Ice Haven, 70° 45′ north latitude, some eighty miles south of Cape Desire.

Faced with the horrible prospect of a winter at that latitude so far beyond the limits of human habitation, the men would not give up hope of escape, though easterly winds, driving more ice into the harbor, reduced their chances every day. On the twenty-seventh the ship was heaved up by pressure, and her rudder smashed. Heemskirk began to land supplies, leaving enough on board to support the crew in case by some lucky chance the ship should be liberated. On September 11 hope was given up and all hands set to work to establish winter quarters. They found driftwood up and down the beach and were much comforted at this sign of God's favor: they now had both building materials and fuel. All labored hard at collecting and stacking it high to prevent it from being buried in drift. On September 26 a gale from the west drove off the ice. In the twilight a black expanse of water was all that could be seen seaward. But the ship lay immovably cemented in the ice of the sheltered harbor.

In late October the house, built of driftwood and roofed over with planks from the ship's lower deck, was declared ready for occupation and all hands moved on shore. Polar bears made journeyings from shore to ship a very dangerous affair. On one occasion three of these animals attacked a supply party when just leaving the ship. Heemskirk and de Veer defended themselves with two halberds carried on the sledge while the rest fled back on board. The bears pursued the unarmed fugitives, giving the two officers a chance of also reaching the ship and sharing in a lively battle on the upper deck. They flung billets of wood at the savage assailants, "and every time we threw they ran after them as a dog useth to doe at a stone that is cast at him." A man was sent below to strike a light (for the matchlocks) and another to fetch pikes. In the meantime a bear, struck on the snout with a halberd, ran away wounded and the others followed. The fatigue party made its way to the hut "and there showed our men what had happened unto us." (One can imagine the eager clamor as each man tried to give his own version of what had occurred.) Early in November with the arrival of the Arctic night the bears moved away to the south leaving foxes the only living creatures around that lonely outpost.

The achievement of those seventeen Dutchmen in wintering through with only two deaths is best measured by comparing their fortunes with those of the crew of Jens Munck at the mouth of the Churchill River, 1619–1620, or of the wintering party of Lieutenant Lassenius east of the Lena Delta, 1735–1736. Their feat remained without parallel until Lieutenant Edward Parry, with all the resources added by more than two centuries of experience and scientific advance, berthed H. M. ships *Hecla* and *Griper* at Melville Island for the winter of 1819–1820. The average of stoutheartedness among the Dutch was no doubt high; they were few in number and too compactly housed for a clique of croakers to form and spread the fatal contagion of grumbling. depression and lassitude. The surgeon, Hans Vos, a worthy teammate to Heemskirk and Barents, devised a sort of Turkish bath which the men passed through in succession and found it "a great means to their health." To supplement their ration of bread, a little over half a pound per diem, they devised traps for foxes and were successful in providing not only a quantity of fresh meat but fur for clothing.

Nonetheless they had more than enough of discomfort and danger. The thunderous cracking of ice seaward disturbed their rest; the chimney was frequently choked with snow; smoke and cold were equally insupportable. As the cold grew more intense the heat of the wood fire was found insufficient; they made a fire of coal, of which the ship carried a small quantity, and, determined to make the most of it, they stopped up the door and chimney and went to bed. Luckily, a man who was weakened by sickness showed signs of asphyxiation; a comrade rushed to open the door and collapsed on the threshold. De Veer brought him round by sprinkling vinegar on his face, and the cold, which had been their deadliest enemy, now served to restore his comrades. A little wine distributed by Heemskirk provided a happy ending to this near tragedy. It was a long time before they again ventured to light a coal fire, though all recognized the cause of the mishap.

As shoes froze "as hard as hornes" they wore moccasins of fox skin over several pairs of socks. Digging firewood out of the drifted snow was a painful task. It was performed in haste by men working in shifts of two, as they could not long endure the cold. Though not so intended, this practice gave every man a daily spell of strenuous effort, and so helped to check, what could not entirely be prevented, the advance of scurvy.

One diary records that December 18 was "a weary time for us to be without the sunne, and to want the greatest comfort that God sendeth unto man here upon the earth." The house was so drifted in that its inmates clambered out as from a cellar. One evening they heard the rustling and scraping of foxes climbing over the rooftop. The more superstitious interpreted this as a bad omen; the question was discussed with some warmth until a sceptic Modernist closed the debate with the good-humored observation that "it was an ille signe because we could not take them, to put them in the pot to roast them, for that had been a very good sign for us."

Though heavily clothed and keeping up great fires they suffered the plague, familiar to the amateur house builder of the North American prairies, of an ice-cold stratum at floor level. To remedy this they sat with their feet on hot stones and scorched their hose before they felt the heat. De Veer asserts that, but for the smell, the socks could have burned through without detection.

Twelfth Night was celebrated with pancakes and a biscuit apiece. In compliance with an earnest request Heemskirk distributed a little wine, which by the cheerfulness it diffused probably did more good than if reserved for an emergency. In mid-January they were cheered by a glow, increasing daily, on the southern skyline. Foxes, which had been a valuable source of meat, were seen less frequently, a sign that the polar bears were returning. On January 25 the sun reappeared, to the confusion of those prophets who had fixed the date of its return without making allowance for refraction.

The scurvy-weakened men now began to go abroad and exercise their joints by running and casting the ball. In early February they were house-bound by a prolonged storm which completely blocked the door and necessitated a man climbing out by the chimney to dig it clear. (A chimney of such dimensions must have been uncommonly drafty.) They killed a bear and used his fat for candle-grease. They rarely ate bear's meat and found, like other travelers in the north, that his liver was extremely poisonous. The more the men moved abroad the more conscious they became of growing weakness. They found it difficult to dig firewood out of the windswept and hardened snow. On March 9 they were cheered by the sight of the Siberian coast away to the south heaved high above the horizon by mirage. On April 3 in clear calm weather those indomitable optimists organized a game of golf, "thereby to stretch our joints, which we sought by all the meanes we could to doe." On the sixth in a storm a bear came down the ramp to the door and tried to force it and then mounted the roof to come down the chimney, luckily without success in either case, for in the hut, which was tightly sealed by frost and overheated, the gunpowder was too moist to ignite.

In the beginning of May the seamen urged Barents to advise

the captain to make a getaway by boat at the earliest opportunity, instead of waiting for the liberation of the ship. The ice was breaking up in the open sea, but in the harbor, sheltered from gales and the disruptive force of the ocean swell, it might be months before the ship found release. The captain and pilot conferred without declaring their decision; when the frightened men grew clamorous Heemskirk dryly "made answer that his own life was as dear unto him as any of ours to us"—they could trust him to make the best provision in his power for lives that were bound up with his own. On May 27 Heemskirk decided on a speedy departure; on the twenty-ninth preparations began. The two boats were dug out of the snow and with immense difficulty—"there wanted no good will in us but only strength"— were dragged to the hut. There they could be repaired and their gunwales raised with convenience and some security against the bears which were continually on the prowl.

The repairs made, they began to move their effects and the boats to the ship, halfway to the open sea, and noted with joy that as the weather grew warmer sledges glided more easily over the snow. "Good will and hope made them stronger" and the hope grew that they might "get out of that wild, desert, irksome, fearful, and cold country." With good sense and discipline prevailing to the last, all hands washed their clothes before embarking. They took up axes, halberds and shovels to clear a road through the chaotic ice from the ship to the open sea. While engaged in this work de Veer was set upon by a bear and barely saved by the surgeon who shot and mortally wounded the savage animal.

When all was ready for departure a history of the expedition written by Barents was enclosed in a bandolier and placed in the chimney of the hut. And so, on 13 June 1597, "committing ourselves to the will and mercy of God, with a west wind and an indifferent open water, we set saile and put to sea." Four of the men including Barents were helpless invalids. Of the remaining eleven the strongest "had but half a man's strength."

With starvation the worst of the perils which hung over them, their best chance was to make contact with Russians or Samoyeds, certain to be found around Vaygach Island, which lay almost directly to the south. But rather than face the uncertainties of an unknown shoreline they elected to take the long way back around Ice Point and down the north and west shores of Novaya Zemlya. They were hardly at sea before ice driven by westerly winds forced them to put in to shore. The stronger men spent their enforced leisure in hunting for birds on the cliffs. On the fifteenth a run of fifty miles brought them to Cape Desire where they passed the night. On the sixteenth they put out again to sea in a wretched drizzle and steered for Ice Point. Heemskirk striving to keep their men's spirits up hailed Barents from the other boat asking how he did? "William Barents made answeare and said, 'Well, God be thanked, and I hope before we get to Wardhouse to be able to goe [run].' Then he spake to me [de Veer] and said, 'Gerrit, are we about the Ice Point? If we be, then I pray you lift me up, for I must view it once againe." They spent that night at sea imprisoned in the ice pack.

On the morning of the seventeenth the boats found themselves in the thick of moving ice in a lead too narrow for maneuvering by oar, and the boats so thumped by small ice pieces "that we thought verily the scutes would burst into a hundred peeces." In this extremity Heemskirk cried out that their only chance was to carry a tow to a large and stable floe nearby and haul the boats into its lee. De Veer, the lightest man in the party, volunteered for this dangerous service, and crawling and clambering from one swaying ice piece to the next he made good his footing on the floe and secured a rope to a hummock on its surface. By this means both boats were hauled into a quiet backwater. The sick were hoisted out and laid on piles of clothing; the boats, "much bruised and crushed with the racking of the ice," were unladen and drawn up on the floe edge. There sprung nails were hammered home and other repairs effected. Some men went ashore to seek eggs, "which the sick men longed for"; they brought back four birds only "not without great danger of our lives between the ice and the firme land, wherein we often fell, and were in no small danger."

Two of the sick for whom these brave fellows ventured so much were fated never to leave the floe on which they had found refuge. On the third day of their detention there "Claes Adrianson began to be extreme sicke, whereby we perceived that he would not live long, and the boateson came into our scute and told us in what case he was, and that he could not long continue alive; whereupon William Barents spake and said, 'I think I shal not live long after him'; and yet we did not judge William Barents to be so sicke, for we sat talking one with the other, and spake of many things, and William Barents read in my card which I had made touching our voiage, and we had some discussion about it; at last he laid away the card and spake unto me, saying, 'Gerrit, give me some drinke'; and he had no sooner drunke but he was taken with so sodaine a qualme, that he turned his eies in his head and died presently, and we had no time to call the maister out of the other scute to speake unto him; and so he died before Claes Adrianson who died shortly after him. The death of William Barents put us in no small discomfort, as being the chiefe guide and onely pilot on whom we reposed our selves next under God; but we could not strive against God, and therefore we must of force be content."

On June 22, after five days detention, it was judged safe to proceed. The storm which had forced them to take refuge behind the floe had pushed much ice in behind them, and the enfeebled crews were compelled to drag their boats hundreds of yards, over wet, ragged and rotten ice before launching again on an open sea. As the weather was not yet warm enough to melt solid drifts, they laid out platters full of snow to melt in the sun, "but all was not enough, so that we were compelled to endure great thirst."

On June 24 they were on the east side of Cape Nassau, 140 miles from Ice Point. On the twenty-fifth a heavy gale drove them to seek safety on the apparently solid ice which lined the shore, but, undermined by the chafing of wind and tide, it gave way plunging both boats into a turbulent sea. De Veer hoisted sail and recovered land, as, after great difficulty, did Heemskirk in the other boat. Some comfort was provided by the discovery on shore of driftwood with which they melted snow for drinking water and dried their bread, which had been soaked in brine. They were held up in that region for more than a week. On July 5 occurred the fifth death, the last of the cruise. Twelve out of seventeen remained alive. On the tenth they were again afloat after a cruel portage over the ice driven against the land, "However it was but folly in us to think of any weariness." Again they were driven inshore and pinned down by ice. Three men landed and found seventy duck eggs. They were at a loss how to carry them to the boats until a resourceful Hollander took off his trousers, secured the legs with cord and so improvised a sack in which the eggs were carried back suspended from a pole. They and their comrades "fared like lords," and with pardonable extravagance topped off the feast with the last of their wine, "whereof every one had three glasses." More good luck was to follow. On July 18 three men mounted a hill and saw that the ice to seaward had dispersed. One more portage, "in good hope that it would be the last time," and they were racing down the coast with strong east and east-northeast winds, traveling seventy miles "betweene every meale-tide," and not forgetting to thank God for this sudden deliverance.

On the twenty-third they were grounded by fog and were embayed for three days as the north wind which would have carried them far on their way was too strong for them to get out of harbor and take advantage of it. On the twenty-sixth they were under way and passed the west end of Matochkin Shar, which they correctly judged to be a strait conducting to the Kara Sea. At St. Lawrence Bay they met Russian fishermen, old friends of Heemskirk and de Veer, who were much surprised at finding in rags the men whom two years before at Vaygach

Island they had seen equipped with such magnificence. The Dutch asked for wine, and received the more wholesome gift of bread and smoked fowl. On July 31 they added to their luck by the discovery on an island of scurvy grass, a remedy of which they were still sadly in need.

Noting that his supply of bread and cheese was nearly exhausted and that the strength of his men unequal to a voyage of much longer continuance, Heemskirk resolved to quit the Novaya Zemlya coast and make a dash by open sea to the Kanin Peninsula whence a short run across the White Sea entrance would bring him to the Lapland shore. On August 3 they launched forth and the next day approached the low tide-swept flats of the Russian mainland. Coasting eastwards they met and hailed a Russian lodia, crying "Kanin Nos? Kanin Nos?" The Russians shook their heads and called back "Pechora." Deprived of Barents' guidance they had set too southerly a course, passed to the left instead of to the right of Kolguyev Island, and struck the coast near the Pechora River some way to the east of the desired peninsula. Their food was now almost exhausted. Some advocated striking inland to find relief at some native settlement. Heemskirk with difficulty restrained his men from feeding on the putrid carcass of a walrus. He was relieved from immediate crisis on the twelfth by purchasing from the crew of a lodia a barrel of fish, to which the Russians, with their accustomed kindness, added cakes and meal. On the fourteenth the boats lost one another in fog, and were not reunited until they reached Lapland.

On the fifteenth a Russian crew advised de Veer that he was still to the east of Kanin Peninsula and warned him by signs not to attempt the crossing of the White Sea in so frail a boat. On the sixteenth in the northwest angle of Cheshkaya Bay de Veer's unbelieving crew questioned Russian boatmen, who confirmed the unwelcome fact that they were still to the east of the desired point. The Russians consented to sell them a little fish only, and the starved and disheartened men were putting off when the captain of the lodia either moved with pity or persuaded by his men, called them back and partly by sale and partly as a gift loaded them with meal, bacon, butter and honey, not for themselves only, but for their comrades in the other boat. He lent them two men in a small boat to pilot them out of the bay and point out the proper course and so sent them off, a little dazed, one imagines, at their good fortune. Two days later they came to the long sought for cape, marked with five crosses, and saw the coast falling away to the south and southeast. To the north up the opposite Lapland shore lay Kildin, Kola and Wardhouse where they hoped to find the Dutch traders who trafficked in those regions. Pausing only to replenish their water supply they shot boldly out into the White Sea. They made the 160-mile traverse in thirty hours and on August 20 found safety in a Lapland fishing village. There they had a taste of civilized comfort. The Russians received them with all imaginable kindness, taking them into their houses, making place for them near the stove where they could dry their clothes, and plying them with all the food they could eat. "We eate our bellies full," says de Veer, "which in long time we had not done." Fortune could not now do enough for the men whom she had so long tormented. Some of the party had gone abroad to gather sorrel grass when two strangers came over the hill and approached the crew of a lodia to barter trousers (of which the Dutch wore two or three pair) for food. Joyfully de Veer recognized them as crewmen of Heemskirk's. The skipper, too, had crossed the White Sea safely and reached the shore a little to the north of de Veer's landfall. After he and his crew had been fed by the hospitable Russians, the two boats departed in company, Heemskirk leaving a gift of money for his host and also an honorarium for the cook, of whose services the Dutch retained a grateful remembrance. On the voyage to Kildin they met and hailed Russian fishermen, crying "Crable pro pal" (Our ship is lost), and received the reassuring answer, "Cool Brabouse crable," from which they gathered that ships of Brabant (the Netherlands) were docked at Kola. On August 25 stormy weather compelled them to put in at Kildin, where Russian fishermen received them with their wonted hospitality. As the departure of the ship from Kola was imminent, they hired a Laplander to guide a seaman to that port with word of their plight. Four days later the Lap returned with a letter promising immediate rescue and expressing the writer's joy at the safety of those whom he had given up for dead. It was signed, "by me, John Cornelison Rijp." To owe their safety to the captain in whose company they had sailed the year before and whom they supposed to have been lost in the Greenland ice seemed to the Dutch a coincidence too extraordinary for belief until Heemskirk produced from his papers a signature which confirmed the identity. Rijp himself put the question beyond doubt by arriving in a hired Russian yawl bringing a cargo of delicacies which de Veer catalogs with loving precision: beer, wine, bread, flesh, salmon, bacon, sugar, "and other things which comforted us much." Taking an affectionate leave of their Russian friends they went on to Kola, where with the permission of the governor, they left their boats in a public place as a memorial to their long voyage and miraculous deliverance. On October 29 they arrived in the Maas to end a cruise which, of all the great voyages of the sixteenth century, is perhaps the most creditable to the stoutheartedness and endurance of those who took part in it.

THE RUSSIANS: COSSACKS ON THE SIBERIAN COAST

The last voyage of Barents was the highwater mark of search for the Northeast Passage by the mercantile powers of western Europe. In 1608 Henry Hudson in the *Hopewell* tried to pass north of Novaya Zemlya and then to force a passage by way of Matochkin Shar. He failed at both points. Dutch whalers became active in the Barents Sea, but geographical initiative passed into Russian hands. Further discovery along the coast arose from developments inland on the great river systems of Siberia. A regular amphibian trade route already existed to the east from Pustozersk at the mouth of the Pechora River, through Yugorsky Shar (Pet Strait), across the Yamal Peninsula by river, lake and portage, and up the Taz River to the settlement at Mangazeya, where at the beginning of the seventeenth century were to be found state officials, merchants, and other folk of varying degrees of disreputability, such as tax evaders and thieves. The respectable part of the population dwelt within a log palisade strengthened with five towers; the homes of the rest were clustered around the walls. A reconstruction of one of these posts closely resembles the Hudson's Bay Company fort of a slightly later date.* By 1676 the trade route to the east extended across to the Yenisey, on whose banks New Mangazeya (Turukhansk) was founded.

At the same time venturesome seamen and traders were pushing up the shoreline beyond the Ob in search of fur, fish and perhaps walrus. Vague reports of the vast northward-reaching

* V. Yu Viese, *Morya Sovetskoi Arktiki* (Moscow, 1940), p. 43.

Peninsula of Taymyr came back. In 1661 the English heard reports of the Khatanga River on the peninsula's east side. Ruins of old dwellings may be seen today past the mouth of the Yenisey. In 1740 the explorers Minin and Strelegov found traders at the mouth of the Pyasina River, but beyond saw no trace of present or former European inhabitants.

Recent evidence proves that at least one ship ventured much further along that forbidding and inhospitable shore. In 1940 the topographical survey party of the *Nord* working on the east side of the Taymyr Peninsula made a find both surprising and of extraordinary interest. On an islet off Faddeya Bay and eighty miles east of Cape Chelyuskin they came upon pots, frying pans, rusty knives and other metal work, besides beads and rotted furs. Excavation of a mound nearby uncovered pewter plates, earrings, and finger rings, old coins to the number of 1171, dating from the latter half of the fifteenth to the beginning of the seventeenth century, crosses of various patterns, some of silver, and other objects of ancient date. The *Nord* spent the winter of 1940–1941 at Cape Faddeya, near the site of their discovery. In April her scientists found in Simsa Bay the ruins of a hut built of driftwood. When the snow had melted close search turned up coins, glass beads, rings, navigational instruments, along with human and other bones. Also a chess set was found and remains of leather shoes with cleated heels. The skeletons included that of a woman or, possibly, a boy.

The finds, studied by the archaeologist A. P. Okladnikov, disproved the theory that the unfortunate voyagers belonged to the Cossacks who were then overrunning Siberia, and later took to the sea east of the Lena River. The wreck was dated about 1618. The Cossacks did not reach the Lena until 1620, and from there their impulse carried them on still farther to the east. Furthermore the ship's furnishings were too sophisticated for Cossacks: they used harsher persuasions than beads to extort furs from the natives. The unfortunate adventurers appear to have achieved the feat which baffled the naval discoverers of the next century by rounding the Yamal Peninsula and Cape Chelyuskin before coming to grief in a cluster of islets of the eastern side of the Peninsula of Taymyr. The voyage supposedly preceded the Imperial decree of 1619 which forbad such intrusions to prevent illegal traffic between western European and Russian traders in the Kara Sea. Prior to disaster the voyagers must have had both skill and luck. Their attempt was not repeated. The conquest and exploration of Siberia was to be achieved not by sailors but by seminomads from the steppes to the north and east of the Black Sea.

In 1581 the outlaw Yermak led a force of less than two thousand Cossacks from this country through the Urals and conquered the region immediately to the east. For this service he received the Imperial pardon, and his conquest was formally embodied in the territories of the tsar. At almost the same time and in a similar manner the Queen of England set the pattern for British imperialism by giving countenance to Yermak's marine counterpart, the freebooter, Sir Francis Drake. The Russians were quicker to develop the colonial theme. Royal officers, the *voevods* Sukin and Myasloi, carrying on the work of Yermak, established the city Tyumen on the site of a captured Tartar stronghold, and pushed on to the Irtysh tributary of the Ob. There they founded Tobolsk which became the administrative centre for Siberia. The vanquished Tartars took refuge on the steppes of Central Asia.

The drive to the east, to which these aggressions were merely preliminary, was facilitated by the magnificent river systems of Siberia. The tributary headwaters of the three great northward-flowing rivers, the Ob, Yenisey and Lena, spreading out east and west, and almost interlocked like the arching branches of a row of trees, afford (with short portages) a trunk line of water transport from the Ural Mountains to the Pacific Ocean, while the three main streams serve as branch lines connecting the great east-west thoroughfare with different sections of the Arctic shore. So, descending the Irtysh, the Cossack invaders reached the Ob and there met the traders of Mangazeya. Commerce and the flag were united. Continuing their eastward march the invaders ascended the Ob and its eastern tributary the Ket, and by a short portage reached a point on the Yenisey where the post Yeniseysk was established. From there ascent of the Upper Tunguska River brought them within easy reach of the Lena headwaters. Far down that river a post was set up at Yakutsk. Below Yakutsk the river Aldan, flowing into the Lena from the east gave passage to the highlands overlooking the Pacific. On that ocean in 1649 the post of Okhotsk was established a little to the south of 60° north latitude. Yermak had forced the passage of the Urals only sixty-eight years before. This gigantic achievement was made possible only by a barbarian independence of transport and supply, coupled with a resolution and fixity of purpose of which few barbarians are capable.

Okhotsk remained an insignificant outpost until the time of Vitus Bering. The main thrust of the invaders was down the Lena to the Arctic shore in quest of furs and the ivory of walrus and mammoth. Undeterred by the perils of the ocean they took to the sea in their flat-bottomed river boats, *kotches*, and reached east to the mouth of the Indigirka. The natives, like the Eskimos of North America, too few and scattered to offer much resistance to bands of strangers, small, but armed and resolute, suffered the twin evils of lust and extortion. The Cossack Robrov of Tobolsk, working from the Lena, discovered the mouth of the river Yana and collected as tribute a great quantity of sables and black furs which he sent to Yeniseysk. He wintered on the Yana and built forts on the Indigirka. In 1642 he traveled west from the Lena to the river Olenek, where "to ensure payment of tribute," he kept native hostages at his fort, in irons day and night, in accordance, it seems, with the instructions of the *voevod* at Yakutsk. There is vague mention of his being sent to the New Siberian Islands off the mouth of the Yana in 1652. In 1653 collection of tribute on the Kolyma River (east again from the Indigirka) caused brawls, and one learns with satisfaction that

Robrov suffered an arrow wound in the head and a broken bone in the back.

There was not in Siberia, as there had been in Spanish America, a pious Las Casas to vindicate the honor of his race by proclaiming, even if in vain, the wrongs of the conquered people. We can only guess that the conduct of Robrov was typical. The Russian government was perhaps as benevolent in intention as that of Spain, but equally powerless to prevent cruelty in the conquered territory, as long as the principle of exploitation was admitted. An Imperial decree required that the Siberian tribute should be collected "mildly, not by severe means," missing the obvious truth that when robbery is authorized, whether it is committed mildly or otherwise, depends less on the agent than on the victim of oppression. No doubt, apart from using native women "for lewd purposes," the Russians employed no more force than the traffic required; and the resistance they encountered seems to have been for the most part passive, or, at the best, irresolute. The Cossack Buza, a contemporary of Robrov, was besieged for some weeks in his fort on the Yana, but appears to have exhausted the hostile natives and collected his quota. The strongest resistance was encountered later from the Chukchi in the extreme east and from the natives of Upper Kamchatka.

The fragmentary record suggests what is, in any case, probable, that on that remote and all but desert coast, the intruders were not always strictly law-abiding among themselves. The Cossack Vagin, who, like Buza, plied his trade between the mouths of the Lena and Kolyma, addressed to the governor of Yakutsk a complaint against the latter. He had purchased for a hundred rubles a Yakut damsel, "for the collection of tribute and as an interpreter," he demurely remarks, "but this Elisei Buza along with Korenev Nifantyev met me at sea, came to me in my boat, beat and maimed me and took away by force the girl I had purchased, my Mandyig, and from the girl they took a coat and ornaments to the value of twenty rubles." This incident, faintly reminiscent of a tale from the *Odyssey*, appears to have done no damage to the credit of Buza, who was later employed in an office of trust by the Yakutsk *voevod*.

The sturdy ruffians working down the Lena from Yakutsk made the coast east to the Kolyma an area of activity for two decades with their quest for ivory and fur. Thereafter the traffic died away. Of the twenty-four voyages in waters east of the Lena Delta in the period 1633–1702 listed by Viese, only five took place after 1656.* The Russians who remained on the coast lived in isolation until the era of scientific discovery. When Pronchishyev went into winter quarters on the Olenek in 1735, he found there twelve Russian (or more probably half-breed) families, who had all but forgotten the ties which bound them to the newcomers. He treated them with kindness, and they gave valuable aid both to Khariton Laptev in that region and to Dmitri Laptev in his explorations farther to the east.

To this period of activity about the mouth of the Kolyma is attributed an exploit of such heroic proportions that, if substantiated, it would rank with the greatest voyages of history. In 1647 one Michael Staduchin was told of the river Pogicha to the eastward which was rich in furs. He endeavored to go there by boat but failed. In the following year in the course of an attempt by land, he crossed the base of the east Siberian peninsula, found the river Anadyr and descended it to its outlet on the Pacific. But his priority in this discovery is disputed. For it is asserted, in 1648, his compatriot, Semen Dezhnev, embarked at the Kolyma, in his flat-bottomed scow, made the long and dangerous voyage to East Cape, turned south through Bering Strait, was driven southeast and after enduring shipwreck reached the mouth of the Anadyr on foot in December, 1648, ahead, so Dezhnev asserts, of Staduchin. Scarcely less extraordinary than the voyage itself is the circumstance that geographers remained ignorant of it, and certainly did not learn the name of its author until 1736 (eight years *after* Bering had sailed up to the East Cape), when the German scholar Muller found Dezhnev's re-

* Ibid. p. 57.

port in the Archives at Yakutsk and published his exploit as assured fact. Muller's prestige was deservedly high; his view was generally accepted until 1886, when the historian Slovtsov "questioned the veracity of Dezhnef and refused to accept Muller's account."* Golder himself subjects the supposed voyage to a searching criticism, arriving at a conclusion wholly unfavorable to the claim of Dezhnev.† L. S. Berg* in a systematic rebuttal disposes to his own satisfaction of what he calls "the unjustified doubts of Golder." The strangest feature of Dezhnev's extraordinary story is that it has stirred so little debate among geographers of all nations.

According to those who support the claim of Dezhnev, six kotches left the Kolyma in June, 1648, commanded by Fedot Alexeev (Viese adds the surname Popov), Imperial agent, Semen Dezhnev and the Cossack Ankudinov. In September they were at Bolshoy Kammennoi Nos (Cape), three boats having been lost en route. At this cape Ankudinov's boat was wrecked and the survivors taken on board the two remaining craft. In an affray with the Chukchi natives Alexeev was wounded. Later the two boats parted; Alexeev's was never seen again. Dezhnev was swept to the southwest past the mouth of the Anadyr and was shipwrecked in October. Ten weeks of wandering on foot brought him and his surviving companions back to the Anadyr in December. Dezhnev's first extant report on his voyage, written some years later, is as follows:

In the year 1648, June 20, I, Semeon, was sent from the Kovima [Kolyma] River to the new river to the Anaduir to find new, non-tribute paying peoples. And in the year 1648, September 20, in going from the Kovima River to sea, at a place where we stopped, the Chukchi in a fight wounded the trader, Fedot Alexeev, and that Fedot was carried out with me to sea, and I do not know where he is,

* F. A. Golder, Russian Expansion on the Pacific, 1641–1850 (Cleveland: Arthur H. Clark, 1914), p. 71n.
† Ibid. pp. 67–96.
* Otkritiye Kamchatki i espedisti Beringa, 1725–1742 (Moscow-Leningrad, 1946), pp. 27 ff.

and I was carried about here and there helplessly until after October 1, and I was thrown up on the beach on the forward end of the Anaduir River. We were in all twenty-five in the kotsh, and we all took to the hills, not knowing which way to go. We were cold and hungry, naked and barefooted, and I, poor Semeon, and my companions went to the Anaduir in exactly ten weeks, reaching that stream low down near the sea. We were unable to catch fish, there was no wood, and on account of hunger we separated. And twelve men went for twenty days up the Anaduir without seeing human beings or reindeer or native trails, and turned back. And when they had come within three days of camp they made a halt. And out of the twenty-five we were left twelve, and we went up the Anaduir in boats and met with the Anauli people.

To go from the Kovima to the Anaduir by sea there is a cape stretching far out into the sea, and not the cape which lies off the Chukchi River. To that cape Michael Staduchin did not come. Opposite that cape are two islands, and on one of these islands live Chukchi, who have pieces of walrus tusk in their lips. That cape lies between north and northeast; and on the Russian side of the cape there is a small river. The Chukchi have a tower of whale bone; and the cape turns around to the Anaduir. In a good run one can go from the cape to the Anaduir in three days and no longer, and to go by land to the river it is no farther, because the Anaduir falls into a bay.

That Dezhnev somehow found his way from the Kolyma to the Anadyr is not questioned. This is his version of how he got there. Three other contemporary reports, to two of which Dezhnev contributes, add nothing in corroboration of his story.

The reader is struck by the vague and fragmentary nature of Dezhnev's account. Berg, conceding this, explains that the voyager did not appreciate the magnitude of his exploit and sought only to prove that he, not Staduchin, was the first to reach the Anadyr. In so pleading he implies that Dezhnev had a business motive for concocting his story, which certainly reads like a not-too-expert improvisation.

Golder is sceptical, firstly on account of the vast distance supposedly traveled with utterly inadequate marine transport. He gives a description of the kotch, which Berg, who has little to say on logistics, leaves unchallenged.

A "kotsh," the kind of boat Deshnef had, was a flat-bottomed decked vessel, about twelve fathoms [72 ft.] long, put together generally without a nail or scrap of iron of any kind, and probably kept together by wooden pegs and leather straps. Buldakof, one of the Siberians, speaks of the ice cutting the twigs of his kotsh. From this statement and hints elsewhere, it would seem that a kotsh was tied together and probably protected on the outside by twigs. A kotsh had a wooden mast and sails of deer skin, which are of little use in damp weather. The chief motive power, therefore, was the paddle. Anchors were made of wood and stone, and cables of leather. This description gives one an idea of the fitness of a kotsh to battle with sea and ice.

One might add that Alexeev and Dezhnev were not skilled professional sailors, but amateurs who used salt water transport merely to improve their mobility as land pirates.

Golder gives the relevant distances as: from the Kolyma to East Cape, 1,115 miles; from East Cape to the Anadyr, 1,045 miles, a total of 2160 miles. What cape is intended by Dezhnev's Bolshoy Kamennoi Nos cannot be proved: the disputant can identify it as he chooses. Berg is confident that it is the East Cape (Cape Dezhnev). Golder argues that Dezhnev's description applies just as well to Shelagskiy Nos (less than half-way from the Kolyma to East Cape), and believes that the expedition was shipwrecked *there,* and that the ten weeks' foot journey in early winter represents the *overland* tramp from Chaunskaya Bay on the Arctic to the Pacific outlet of the Anadyr. Berg finds this hypothesis "utterly senseless. . . . Kallinikov did make this journey with dogs in 1909, but it cost him much labor and demanded three weeks for its completion."[*] But the Soviet historian forgets or ignores the objection of Golder that the obstacles to a voyage from the Kolyma to East Cape were even greater. The American cites no fewer than six attempts in the period 1649–1787 (including the strenuous efforts of Dmitri Laptevm 1740–1742), none of which succeeded in getting past Shelagskiy Nos.[†]

[*] Ibid., p. 37.
[†] Golder, *Russian Expansion,* p. 95.

The identification of the Bolshoy Kamennoi Nos with East Cape is not much helped by Dezhnev's statement that "in a good run one can go from the cape to the Anaduir in three days and no longer"—an utter impossibility. Perhaps he intends not East Cape but its echo to the south, Cape Chukotskiy. From there the distance is "not less than 500 miles (Golder); about 460 miles" (Berg); "a distance far too great for any kotsh to make in 'three days' and 'no longer'" (Golder); "at a speed of 6 knots it can be traversed in three days" (Berg). Perhaps seven or eight knots would be nearer the mark, unless Dezhnev knew his exact course and steered it with remarkable precision. In any case a breeze strong enough to drive his unhandy scow through the ocean swell at six knots would have swamped it, or at least, so have drenched its deerskin sails as to make them useless. But, Berg pleads, Dezhnev "was pursued all the time by abominable weather from Cape Dezhnev until cast upon the beach." His "three days" are not observed time but an informed estimate. This is to replace an improbable hypothesis by a patent absurdity, for we are in effect asked to believe that while being driven hundreds of miles by a tempest, without, presumably, compass or other navigational aid, and subsequently wandering on land for ten weeks, "not knowing which way to go," Dezhnev had kept his dead reckoning so exact that he could confidently reckon the distance across Holy Cross Bay as three days and "not a bit more," "bolshe net." As two knots was probably good going for a kotch, one is tempted to believe that he is alluding to the crossing of Chaunskaya Bay, that his shipwreck occurred, as Golder suggets, at Shelagskiy Nos, and that his trek of ten weeks does represent the foot journey from the Kolyma to the Anadyr.

Golder makes the telling point that, although Dezhnev resided at Yakutsk, "the gathering spot of all Siberians and Arctic navigators" until 1671, he was "totally ignored by his contemporaries" who took no notice of his alleged exploit, "not even to deny it." To refute this Berg invokes "a cloud of witnesses" so voluminous and so unconvincing, as to prove to what extremities the learned Russian is reduced to make even a pretense of countering the arguments of Golder.

(1) A sketch map produced at Tobolsk by the *voevod* Peter Godunov at the order of the tsar, Alexei Mihailovich, shows a water route from the Lena to the Amur. The insertion of the River Kamchatka, Berg says, "could only be the result of the journeys of Dezhnev, Staduchin and their successors on the Anadir." This has no bearing on the point: no one disputes that Dezhnev was on the Anadyr; the question is how he got there. And the Russian, a very inept debater, spoils his intended point by naming with Dezhnev, Staduchin, who admittedly came to the Anadyr by land.

(2) A similar map, based on the first, shows East Cape, but damages the case for Dezhnev by calling it the "impassable cape."

(3) Nikolai Spapharie, Russian ambassador to China, 1677–1678 published a description in which he visualized Eastern Siberia as terminating in a long mountainous cape.

(4) A Catholic father, Urei Krizhanich, after fifteen years of exile at Tobolsk, published in 1680 his "Historia de Sibiria" in which he stated that Siberia and China are washed by the same continuous ocean.

(5) In 1697 the Dutch cartographer Witsen produced a map showing the East Cape.

(6) In 1706 the French geographer, Guillaume DeLisle produced a map of Northeast Asia, perhaps based on Witsen, showing the northeastern extremity of Asia as a narrow promontory formed by a chain of mountains. He adds the note: "On ne sait pas où se termine cette chaine de montagnes, et si elle ne va pas joindre quelque autre Continent." That is, he is not sure that America and Asia are disconnected, and is either ignorant of Dezhnev's pretensions or thinks them unworthy of notice.

(7) Strahlenberg, an officer in the Swedish army captured at Poltava and for thirteen years (1709–1722) held prisoner in Siberia, published a map of the Russian Empire with this note op-

posite the mouth of the Indigirka: "From this point Russians traversing the sea through mountains of ice which north winds drive against the shore and south winds carry back seaward, made their way with enormous trouble and danger to the region of Kamchatka." In all the examples cited by Berg to prove that Dezhnev's voyage was known prior to Muller's discovery this is the nearest we come to a direct allusion to the cruise. But it is far from convincing. Strahlenberg's notes seem to refer to a series of voyages, not to a single one; he neither names Dezhnev nor appears to refer to him. Dezhnev can have had little trouble with ice, as Muller admits, in order to make the voyage in one season.

(8) In 1725 Vitus Bering wrote from Yeniseisk to the government to point out that according to the evidence of maps, supported by popular opinion, it was possible to sail from the Kolyma to the Anadyr. The maps can have inspired little confidence when Bering draws them to the attention of his superiors and adds as confirmation, "the folk out there say so"; ("zhiteli ckazivayt") in the original.

(9) The myth that Fedot Alexeev, who parted from Dezhnev in a storm near the Bolshoy Kamennoi Nos, made his way to the southern extremity of Kamchatka is barely relevant and not worth repeating.

(10) Berg also records that in 1711 Chukchi natives informed one Popov that long before Russians had come to the East Cape in kotches. If confirmed this would be the strongest evidence which his painstaking researches have brought to light. For the rest his attempt to prove that Dezhnev's voyage was an accepted fact prior to 1736 is wholly futile. He furnishes abundant proof that in the last half of the seventeenth and first quarter of the eighteenth century the general trend of the coast from East Cape to the River Amur was known. But his witnesses discredit the very thesis which they are quoted to support. Not one of them has heard of Dezhnev. Berg admits this. And his contention that late seventeenth-century beliefs regarding Bering Strait and East

Asian coastal trends must be derived from the voyage of Dezhnev are absurdly mistaken.

As soon as it was established that Columbus' discovery was not the true Indies, geographers began to speculate on the probability of America being a continent divided by ocean from the land of Marco Polo. This view was illustrated in Schoner's map of 1523. Late in the century the English traveler, Sir Humphrey Gilbert, wrote a learned treatise to prove that the continents were divided and that China could be reached from Europe by sailing through the Northwest Passage. He supports his theory by references to Plato and Aristotle and by citing cosmographers of China "by whose experiences America is prooved to be separate from those parts of Asia *directly opposite the same.*" Gilbert's assertion that the Northwest Passage "hath been sayled through" in ancient times is a reminder of how readily the story of a mythical voyage can obtain credence. Muller and he were equally uncritical. Mercator's map of 1587 gives a fairly accurate picture of the east coast of Asia, including a recognizable delineation (perhaps furnished by Jesuit missionaries in Japan) of that Kamchatka which Berg would have us believe was first made known to geographers by the researches of "Dezhnev, Staduchin and their successors" more than half-a-century later. Muller's acceptance of the Dezhnev story and the subsequent acquiescence of scholars in general should not be given undue weight. The German scholar would, with perfect sincerity, be inclined to make the most of his find in the Yakutsk Archives, and there are few western scholars with the capacity or the will to scrutinize the evidence. The case of the Piltdown man illustrates the vitality of an imposture when it has once been endorsed by respectable authority.

The promontory which Muller identifies with Dezhnev's Bolshoy Kamennoi Nos—next to the Horn and the Cape of Good Hope the most conspicuous on the globe—was given an accurate survey in 1778 by Captain James Cook, who named it the East Cape. With horrible gaucherie Russian geographers have altered this to Cape Dezhnev. Now the name given by Cook is so ap-

propriate to the eastern extremity of the Eurasian land mass—
the almost exact antipodes of Greenwich in terms of longitude,
the place where annually the New Year is first welcomed—that
to supplant it by a name of doubtful and somewhat disreputable
provenance is an affront to world geography and a slur on the
great navigator who first gave the cape an intelligible shape. If
South Africa in a fit of chauvinism should propose to rechristen
the Cape of Good Hope after some obscure *voortrekker* who
made his living by plundering the aboriginal inhabitants in the
guise of a tax collector, mapmakers would reject the change
with contempt. East Cape should be redeemed from a similar
indignity.

Not without interest in this connection are the comments of
Captain James Burney, who was both a scholar and a practical
seaman, and had shared in the charting of the shores of Bering
Strait by Cook: "Some memorials of the early expeditions of the
Russians, which were preserved at Jakutzk, and which came into
the hands of Professor Muller, have furnished almost all that is
known of the voyages of Deschnew and his companions in 1648.
Mr. Muller's history of it is written in language plain and im-
pressive; but an inadvertence in the method of relating the order
of events, in its commencement, has rendered his narrative perplexed; and he adopted, upon presumptive circumstances, a view
of the subject which can only be established by absolute and in-
disputable proof."[*]

Burney concludes from a study of Bering's instructions on his
first voyage: "All this shows, that neither intelligence nor credited report had then reached *Europe*, of a navigation having been
performed round a Northeast promontory of *Asia*."[†] If at any
time it had become known or received as a fact in Siberia, that
Deschnew (a remarkable character) [that is, a well-known per-
sonality, whose claim would have commanded notice, if not

---

[*] Captain James Burney, F.R.S., *Chronological History of North-Eastern Voyages of Discovery, etc.* London, 1819, p. 63.
[†] Ibid., p. 120.

respect] "had, in the year 1648, passed wholly by sea round the Tschukzki coast, it is not credible that a circumstance so extraordinary should have so speedily and entirely passed out of men's minds, as for no traces of it to have been discovered by the enquiries and examinations which were made, by order of the Russian government, about the year 1710; and that nothing of traditional report should have remained concerning it."*

* Ibid., p. 127.

BERING, AND THE GREAT NORTHERN EXPEDITION

The Cossack occupation of northeast Asia and its subsequent extension into Kamchatka stirred up in Russia the interest in the Northeast Passage which had died away among the older maritime powers of Europe. In 1713 Fedor Saltikov, an enlightened counsellor of the tsar, Peter the Great, proposed a scheme for the exploration of the Arctic route to "China and other islands," and presented a detailed plan for the charting of the Siberian north shore (which was to be implemented twenty years later). In the expansive mood generated by the victory over Sweden, the Russian statesman seems ready to challenge the commercial supremacy of England and Holland, many of whose traders, he points out, perished from fever and bad provisions on the long voyage around the Cape of Good Hope to the East Indies. He expected the Siberian sea to prove a better channel for the trade, and with a hustling optimism worthy of an old-time promoter in the American West he urged the establishment of a fortress on the Vaygach Strait to levy tolls on the foreign commerce traveling to the Orient by that route. Peter attended the more readily to these counsels as he was being urged in the same direction by men whose interest in the matter was purely scientific. On his visit to Paris in 1717 he had conferred with the geographer, Guillaume DeLisle, on the question of the continuity of Asia with America, and had corresponded with the philosopher Leibniz on the same topic.

His first moves were tentative and ineffective. In 1719 two

members of the Naval Academy, Evreinov and Luzhin, were sent to make a somewhat irrelevant survey in Kamchatka and on the Kuril Islands. In 1720 the geodesist Chikachev, and the geographer Muller embarked on a voyage of discovery east from the Ob. They were thwarted by ice. Ever quick to learn, Peter concluded that the task was one for salt water scientists. Judging reasonably that the vast undertaking of charting the Siberian north shore had no practical justification until the strait between Asia and America was certainly known to exist, he wrote out with his own hand instructions for the "Siberian expedition," dated 26 January, 1725, thus quoted by Golder:*

1. To build in Kamchatka or in some other place one or two decked boats.
2. To sail on these boats along the shore which runs to the north and which (since its limits are unknown) seems to be a part of the American coast.
3. To determine where it joins with America. To sail to some settlement under European jurisdiction, and if a European ship should be met with learn from her the name of the coast and take it down in writing, make a landing, obtain detailed information, draw a chart and bring it here.

Foreign observers had criticized the orders given to Chikachev as too vague. Peter made no such error in the instructions prepared for Vitus Bering.

Viese finds in these orders "strong confirmation for the belief that the existence of Bering Strait, discovered by Popov and Dezhnev, was well known in the time of Peter [from Strahlenberg's map]. Consequently Bering's expedition was dispatched not only with the aim of confirming the existence of a strait between Asia and America, but had some more remote object of a practical nature, evidently disguised for the sake of secrecy."†

An alien studying clauses 2 and 3 of the above orders could hardly avoid drawing a conclusion directly opposite to that of

---

* Golder, *Russian Expansion*, p. 134.
† *Morya Sovets*, pp. 61–62.

the Soviet patriot, Viese namely, that Peter knew nothing of Bering Strait and was inclined to believe Asia joined to America. Viese himself perceives that it would have been ridiculous to mount a long and costly expedition merely to confirm what, on his showing, was already known from Strahlenberg's map; so he postulates for Bering's voyage "some more remote object . . . disguised for the sake of secrecy." But this is to explain away a gross absurdity by one even more glaring. Bering knew nothing of "some more remote object": he believed that he had fulfilled his mission by proving the existence of a strait between the continents; and an expedition with a purpose kept secret even from the officer who is to carry it out is a paradox of which not even the imperious Peter was capable.

Even when he drew up these orders the emperor was dying. Though eager personally to supervise the preparations for the expedition, he wrote to Admiral Apraxin, "My cursed health compels me to sit at home." But still zealous for credit of the empire he had founded he added: "To protect our fatherland from its enemies we must strive to add glory to the empire. May we be more fortunate in the discovery of this route than the Dutch and English who have often tried to open up the shores of America." He died a few days later.

The command of the expedition was entrusted to Vitus Bering, a Dane, some forty-five years of age, who had served twenty years in the Russian Navy. His principal officers were lieutenants A. Chirikov and M. Spanberg, and Midshipman P. Chaplin. Peter had written: "It is very necessary to engage as Pilot and Second Pilot men who have been in North America," but sailors with this qualification could not be obtained.

Bering's feat of carrying a numerous body of men bearing ironware, and all the rigging for oceangoing vessels which the remote coastal station of Okhotsk could not supply was made possible by the rivers of Siberia which provide a fairly direct and almost unbroken system of waterways from European Russia to Yakutsk. Arriving at Tobolsk, March 16, 1725, Bering and his party boated down the Irtysh to the Ob, ascended the latter to its junction with the Ket, followed the Ket to the highlands, portaged seventy versts to the Yenisey and went up it and its tributaries, the Tunguska and the Ilima to Ilimek, reaching it September 29. There the expedition wintered and built boats. In the spring they made the short portage to the Lena and boated down to Yakutsk, separated by 1,000 versts of a road which crosses a low range of mountains from the waters of the Pacific at Okhotsk.

At Yakutsk Bering broke up the expedition into three detachments. He himself set out over the hills with packhorses and reached Okhotsk October 1. Spanberg, appointed to bring on the heavier equipment as far as possible by water, had a rough passage. He boated down the Lena to its junction with the Aldan and up the latter until stopped by shallows. Cruel hardships were suffered in freighting over the height of land. Many packhorses died and the starving carriers were barely kept alive on their flesh. By adopting a system of relays Spanberg was able to deliver his men and some of his freight at Okhotsk by January 12, 1726. The more fortunate Chirikov, who had been left to winter at Yakutsk, brought his detachment over safely in July.

Okhotsk was a tiny village on the east shore of a gulf separated from the open Pacific by the Kamchatka Peninsula and its extension in the chain of the Kuril Islands. Rather than risk the uncharted course around Cape Lopatka the explorers adopted the safer but incredibly clumsy procedure of building a ship at Okhotsk, sailing themselves to Bolsheretsk on the Kamchatkan west shore, and freighting across the peninsula to Nizhne Kamchatka on the Pacific shore. In this last stage they spared the aborigines no more than they did themselves: "To carry the expedition's freight over the 900-kilometer journey native dogs were used, a considerable proportion of which perished with the result that the natives were left without the means of transport."* In the spring the *St. Gabriel* was built, brig-rigged, 60' by 27' 6", and manned with a crew of forty-four supplied for one year. She sailed on her northern mission on July 13, 1728.

Though often hampered by fog and shouldered to the northeast by the trend of the shoreline, the little ship made good headway and by August 8 was past Holy Cross Bay and near the crisis of the voyage. Chukchi were spoken to but did not prove informative. Chukchotski Nos, the southern of the two horns which from the east end of Eurasia, was rounded, and on the eleventh, the ship was in the strait which divides St. Lawrence Island from the continent of Asia. At 65° 30' north latitude, more than halfway from Chukchotski Nos to East Cape, Bering called a council of war and gave it as his opinion that they had accomplished their mission by proving the continents divided. Chirikov, while agreeing with his chief that, if they had shown Asia to be separate from America, they were absolved from the duty of visiting the latter continent, held that "we cannot know with certainty whether America is really separated from Asia unless we touch at the mouth of the Koluima" [Strahlenberg's map and note, which Berg and Viese insist was known to the explorers, seems to have been disregarded here] "or until they encountered the polar ice." He was therefore in favor of persevering, and of wintering, if the need arose, at Chukchotski Nos, rather than return with an inconclusive report. Spanberg advocated an early return to Kamchatka. Continuing their course they passed the true "landes end," later to be named East Cape by Cook—"a Peninsula of considerable height, ... it shew[s] a steep rocky clift next the Sea and off the very point are some rocks like Spires." Bering, for his part, held a northerly course for another seventy miles and saw the coast of Asia falling away to the west. In poor visibility he was not aware that the American shore lay not many miles distant on the starboard beam. On August 16, in 67° 18' north latitude, 176° 53' west longitude, Bering declared their

* Ibid., p. 62.

mission fulfilled and put the ship about. Diomede Island was sighted for the first time on the way back. On September 7 the *St. Gabriel* again cast anchor in the Kamchatka River.

In the summer of 1729 Bering made a voyage directly to the east in search of America, as a gesture, one suspects, rather than with well-grounded hope of success, for he put back when 200 kilometers out, and following the course which he had, at the cost of much labor, rejected two years before, he sailed around Cape Lopatka, the southern extremity of Kamchatka, to Bolsheretsk and thence to Okhotsk. In March, 1730 he was back in St. Petersburg, five years after his departure for the Pacific.

An immediate aftermath of Bering's first voyage was the crossing of the strait he had discovered. In 1726 the Cossack Shestakov reported a "Great Land" north of the Kolyma River, one of those mythical regions which bloomed so plentifully in the romantic period of geographical discovery. The authorities granted him a commission to make discoveries in northeast Siberia and to "pacify the turbulent Chukchee." When he was killed by the latter, his helpers, the Second Pilot Federov and the geodesist Gvozdev, secured possession of the *St. Gabriel* and sailed to Bering Strait, where, mistrusting the visionary land to the north, they went east and made landfall on Cook's Cape Prince of Wales on the Alaska shore, establishing a Russian link with the North American continent. Thus far had the enterprise, aided by the rivers of Siberia, carried them from the eastern frontier of Europe. At that date western Europeans, laboriously advancing from the opposite quarter, had barely reached Lake Winnipeg.

Bering's reception at the Russian capital by the Admiralty and the savants, native and foreign, who made up the Academy of Sciences, though not unfriendly, was not cordial. The fact was that few great explorers have felt less of a vocation than he. He was not of the old school of flamboyant patriots who sought to discover and claim for their sovereign any new land not already in the possession of a "Christian Prince." And the class of trained navigators who, if not themselves versed in science, were zealous in promoting it, had not yet emerged (Bougainville and Cook were still in their childhood). The old Dane was a brave and experienced master mariner who executed his orders with all the thoroughness and exactness which conditions permitted but without any imaginative overtones. It was probably for the sake of his own credit that Bering advanced a proposal for the exploration by sea of the Siberian north shore already advocated by Saltikov.

The scheme was readily accepted, for the discovery of Bering Strait seemed to offer a real prospect for the opening up to mercantile interests of a more convenient route to Kamchatka than the interminable overland journey across all Siberia. In 1732 a decree authorised the attempt and gave the overall command to Bering. In 1733 the participating units began to move toward their respective points of departure.

The river pattern of Siberia made possible the division of the north shore into a number of sections all fairly accessible by way of the Ob, the Yenisey or the Lena. The flowing rivers of the flat north Siberian terrain permitted the construction of boats up to seventy feet in length at settlements with some industrial facilities hundreds of miles upstream from the salt water where they were to operate. This was a convenience which the British and American discoverers of the next century, who made their approach to the Arctic by way of turbulent streams unnavigable by anything larger than canoe or open boat, might well have envied. The first discovery unit, based in Archangel, was to explore east from the Pechora River to the Gulf of the Ob. The second, starting from Tobolsk, was to go down the Ob and chart eastward to the Taymyr Peninsula or the Khatanga River. The third, based on Yakutsk, was to descend to the Lena delta and work westward to the Taymyr or beyond to the Pyasina River. The coasts assigned to detachments (two) and (three) overlapped a great deal to provide for one party or the other failing to complete its prescribed task. The fourth unit, also based on

Yakutsk, was to go down the Lena and work east to Bering Strait, along a coastline in extent hardly less than the collective assignment of the other units. In addition ships based on Okhotsk were to cross the Pacific to the northwest shore of America. Bering took direct charge of this enterprise. The seaborne expedition under Spanberg was to visit the Kuril Islands and Japan, and provision was also made for inland exploration in North Siberia.

The eastern detachment under lieutenants Muravyev and Pavlov, with two kotches, fifty-four feet in length, bearing fifty-one men including two mineralogists, left the North Dvina on July 21, 1734. They passed Yugor Strait on the 25 July and at the end of the month were near the Yamal Peninsula. They coasted up this to 72° 35′ north latitude and on the brink of success turned back to winter at Pustozersk. In the next season conditions were less favorable. They did not pass Yugorski Strait until August 17. Nonetheless they bettered the record of the previous year, and ascended the west side of the Yamal Peninsula, Muravyev going to 73° 04′ north latitude, Pavlov to 73° 11′. The latter taunted his chief with incompetence. They again wintered on the Pechora where both the local inhabitants and members of the expedition complained of the behavior of the two officers. They were recalled for trial by the Admiralty College and in a comprehensive verdict found guilty of "disorderly, careless, lazy and stupid conduct," and degraded to the rank of seamen.

The harsh but able Lieutenant Malygin, who succeeded to the command, blamed the misfortunes of his predecessors on the unhandiness of their boats, and obtained replacements. Owing to a very bad season he got no farther than the Kara River. But instead of returning to the more congenial Pechora, he wintered on the spot and starting from there in 1737 he forced his way around the Yamal Peninsula and went up the Ob gulf to reach the river mouth on September 22. Thence he returned by land to make his report to the Admiralty. His second-in-command, Lieutenant Skuratov, charged with bringing the crews home, was icebound off the Kara River and did not reach Archangel until 1739. A comparatively small operation in terms of distance had occupied two crews for six seasons. These difficulties raised the question of how well fitted the kotches were for this service. Malygin had found Muravyev's boats unhandy in tacking, a grievous handicap in ice navigation. The coast as far east as Yugorski Strait was already familiar to sailors both native and foreign, but Muravyev and Malygin may be credited with the charting of the shoreline around the Yamal Peninsula and up the west side of Ob Gulf.

Lieutenant Dmitri Ovtsin had been appointed to chart the coast from the Ob to the Yenisey and thence to the dimly known Taymyr Peninsula. His boat was built at Tobolsk, 600 miles up the Ob. Setting out on May 14, 1734, he set up his base at Obdorsk, near tidewater on the Ob and carried his coastal survey to 70° 4′ north latitude, still well within Ob Gulf. When scurvy broke out in the Obdorsk winter quarters he moved his men upstream to what he deemed the healthier air of Berezov. In 1735 he was ice-bound in Ob Gulf at 68° 40′ north latitude. Four men, including the mineralogist Medvedev, died of scurvy. Ovtsin gave up early, took his crew upstream to Tobolsk and made the winter journey to St. Petersburg to request better officers and equipment. In 1736 with his old boat he failed at 72° 40′ to get out of the Gulf of Ob. In 1737 with new boats and Pilot Fedor Minin as second-in-command he emerged from Ob Gulf to 74°, rounded to the east, sailed up the Yenisey estuary and on to Turukhansk.

Ovtsin had not yet completed his mission which required a further survey to the Taymyr Peninsula, but he was the only one of the officers originally appointed to achieve anything considerable before being replaced. He was the first to go by sea from the Ob to the Yenisey, for the traders had used the short overland route. His services received very shabby acknowledgment. For courtesy shown to the political exile, Prince Dolgoruki, at Beresov he was arrested, reduced to the rank of seaman and sent to serve on Bering's transoceanic voyage.

The completion of Ovtsin's assignment was entrusted to Minin along with Second Pilot Dmitri Strelegov and the mineralogist A. Leskin. In his first season in 1738 Minin charted the east side of the Yenisey Gulf by ship to its northeast angle, Cape Northeast (Severo-Vostok) at 73° 7′ north latitude. There he left a memorial carved on a wooden slab which was recovered with the inscription still clearly legible in 1922. In 1739 a late start accomplished nothing. In 1740 Minin sailed northeast up the coast to 75°. Strelegov took to the land with dog sledge and extended the chart from the Yenisey past the mouth of the river Pyasina to 75° 20′ before snow-blindness compelled him to turn back. But by reverting to land travel the Russians made a virtual surrender of the commercial objective of the expedition. They still pursued its geographical aims with tenacity, but no more progress was made from the Yenisey. Minin was charged by Strelegov with dishonesty, cruelty and drunkenness; he retorted with complaints of insubordination, but was convicted and degraded from his rank for two years.

Lieutenant Pronchishyev, with Yakutsk as his base, had been assigned the duty of carrying a survey west from the Lena Delta and linking it with Ovtsin's explorations from the Yenisey. One or the other was to surmount the obstacle of the northward jutting Taymyr Peninsula. Madame Pronchishyev accompanied the expedition.

The boats set out on 30 June 1735, a late start for Yakutsk was more than 600 miles from the sea. By August 2 they were in the Lena Delta and, emerging by a westerly exit, followed the coast to the mouth of the Olenek River where low temperatures and a frozen sea compelled them to set up winter quarters on September 5. Nearby was an isolated colony of twelve families of Russian traders and hunters. Pronchishyev did not molest these poor folk by billeting—his party of fifty was too numerous to profit much from so small a settlement. He built barracks and brought his party through the winter fit and ready for a fresh thrust at their objective.

As commonly happens, the ice of the estuary was late in breaking up. The expedition did not get away until the middle of August. Its commander was already stricken with scurvy. They reached the mouth of the Anabar where inhabitants told the geodesist Chekin of mineral deposits upstream. He procured samples which—passed on through Bering to Professor Gmelin —were declared worthless.

Past the Anabar the ships encountered ice and proceeded in constant danger. On August 24 the mouth of the Khatanga was noted as affording a good wintering place. Advancing north and then west, they rounded the northeast shoulder of the Taymyr Peninsula. Pronchishyev had all but attained the prize for which he was laying down his life. Not many miles ahead, reaching out beyond 77° north latitude, lay the northernmost point in Eurasia. But he was barred from it by ice "to which no end could be seen," and on the ice the wondering sailors noted bears "like cattle." A council of war resolved to return to the Olenek. Contrary winds drove them back from their former anchorage. Pronchishyev died on his ship, September 9, and his wife died two weeks later on land. They were a brave but pathetic couple. The command devolved on the Pilot Chelyuskin, who went back to Yakutsk to notify the Admiralty College of all that had happened and to await orders.

Pronchishyev had so nearly surmounted what had been recognized as the chief obstacle on the north Siberian route that the authorities did not hesitate to order a renewal of the attempt. They appointed Lieutenant Khariton Laptev to replace Pronchishyev and complied with all the requisitions which that officer, a man of prudence and foresight, presented for their approval. Storehouses were set up and fishing stations manned on the Anabar, Khatanga and Taymyr Rivers, providing advanced supply bases and a measure of insurance against the plague of scurvy. In June 1739 he sailed from Yakutsk with a crew of forty and the pilots Chelyuskin and Chekin as his officers. On August 4 he was out of the Lena Delta and off the Olenek. On the eighth he discovered a bay to which he gave the Scandinavian name of Nordvik. Passing the Khatanga and wheeling to the north with the bend in the shoreline, he coasted along the east and north sides of the Taymyr Peninsula to Cape Faddeya on September 1, where he placed a "mark," rediscovered 180 years later by Amundsen. Here he encountered "sea beasts," resembling fish, white with black noses and called by the local inhabitants, "Beluga." Here also, like Pronchishyev, he was brought to a dead stop by ice. He returned to the Khatanga and found a winter berth at the mouth of its tributary, the Bludnoi.

Khariton Laptev, who of all the detachment leaders was to fulfill his commission with the highest degree of completeness and economy, owes his success in part certainly to the experienced aid of Chelyuskin and Chekin, but also in great measure to his own foresight and resource. His orders required him to extend his survey east to the Pyasina River to provide against the failure of the Ob detachment to reach the Taymyr Peninsula. In early April, 1740, while still hoping to accomplish this by sea, he insured himself against failure by sending Chekin with dog sledge and reindeer to start a survey westward from the mouth of the Taymyr River, and the boatswain Medvedev, to follow the coast east from the Pyasina until he joined with Chekin. Chekin rounded Taymyr Lake, descended the river to its mouth and laid down one hundred versts of shoreline, when he ran out of provisions and got back to his base on May 28 "in the direst necessity." Medvedev also returned after mapping some forty kilometers of his assignment. These journeys were more profitable in travel experience than in positive achievement. Laptev's response was prompt. He recruited two parties of local inhabitants and sent them to fish and set provision depots both on Lake Taymyr and at the river's mouth.

With the coming of spring the season for travel by sledge on land and shore ice ended and that of discovery by ship approached. In the summer of 1740 Laptev renewed the attempt to complete his task by sea. On the east side of Taymyr Peninsula at 75° 15′ north latitude his ship ran into ice and was fatally nipped. The crew reached land after a fifteen mile journey over the ice and began the weary task of freighting supplies from the wreck to the shore. Disorders—to which a shipwrecked crew is prone—appeared, aggravated by exhaustion and the fears natural to men on an icebound and barren shore, some three hundred miles from their Khatanga base. Laptev maintained order with prompt and judicious severity: he had supplies all landed by September 11 and was back at his base by the end of October. He now had no option but to carry out his work on foot. On March 28, 1741, Chelyuskin was sent with sledges to explore from the Pyasina River, the western boundary of Laptev's section, east to the Taymyr River. Chekin was sent to carry up the east side of the Taymyr Peninsula the survey twice attempted by ship. With a touch of strategic insight Laptev sent his spare men not up the Lena whence they had come, but overland to Turukhansk on the Yenisey, presumably as well supplied a post as Yakutsk, and much nearer the theater of operations. He himself went down the Taymyr River and followed the coast to the north.

On the east side of the peninsula Chekin reached 76° 35′, some distance short of the latitude attained on shipboard, and turned back with snow blindness. Laptev went up the west side to 76° 42,′ and being also stricken with snow blindness, returned to his depot on the River Taymyr. He then set out southwest along the coast and met Chelyuskin in his ascent from the west. From the Pyasina in the west and the Lena in the east the dogged assailants had pushed their way far up the sides of the Arctic stronghold. Its summit, which they called then Cape Northeast, remained stubbornly inaccessible. The leathery Chelyuskin was given the task of overcoming this last obstacle. He quitted the party's winter quarters at Turukhansk at the end of March,

1742, and in order not to arrive at the Taymyr depots until his sledge-borne supplies were exhausted, he crossed to the Khatanga and commenced a counterclockwise circuit of the peninsula. On May 14 he was at Cape Faddeya, the farthest of previous surveys; on the twentieth he rounded the northernmost point of the Eurasian land mass, the projection from the Taymyr Peninsula thenceforth to be known as Cape Chelyuskin. On the west side he linked up his survey with Laptev's of the previous season and reached the River Taymyr, where he met two men sent by Laptev to assist him, closely followed by Laptev himself. The ready and almost superfluous help given Chelyuskin on his last journey shows how readily Laptev and his men had adjusted to the techniques of travel over the Siberian tundra.

Chelyuskin has suffered the fate, to which every explorer obliged to report an important discovery without authoritative corroboration is liable of having his results questioned. Ferdinand Wrangell in his *Narrative of an Expedition to the Polar Sea* finds his observations superficial: Baer thinks that "from dislike of his task he forged results."* Others accept his findings as genuine, including Nordenskiold, who, as the next man to visit Cape Chelyuskin 130 years later, may be assumed to have checked his observations with care. Certainly the 77° 34′ north latitude which he assigns to the cape is a satisfactory approximation to the 77° 42′ of the modern Atlas.

Exploration from the Lena east had been entrusted to Lieutenant Peter Lassenius, a Dane newly enlisted in the Russian service. He cleared the Lena Delta August 18, 1735, soon encountered ice and on August 27 sought winter quarters in Borkhaya Bay. Scurvy attacked the party with extreme severity; Lassenius died on December 30, and of his party of forty-five only eleven outlived the winter. On learning of this disaster Bering, then at Yakutsk, sent Lieutenant Dmitri Laptev, half-brother of Khariton, with forty-three men to carry on the work. Laptev, naturally making a late start in 1736, met ice at Borkhaya Cape

* Ibid., p. 74.

and returned to the Lena. With most of his party suffering from scurvy he went up to Yakutsk next summer and not finding Bering, who had gone forward to Okhotsk, he traveled to St. Petersburg to report the impossibility of "sailing from the Lena to the Kolyma and thence to Kamchatka."

Viese is critical of Dmitri Laptev for "showing a lack of perseverance and giving up on his first encounter with the ice." (Along with James Cook and Joseph Billings, Laptev demonstrated the extreme difficulty of a voyage which Dezhnev is supposed to have accomplished with nonchalant ease.) The Admiralty College judged the young officer less harshly and, it would seem from the sequel, more wisely. They encouraged him to make another effort by sea, failing which, he was to carry out the survey from the shore. And so, setting forth again from the Lena in the summer of 1739, Dmitri Laptev met the ice at Borkhaya Cape, got through "with much anxiety and alarm" and on August 25 was off Sviatoi Nos in Laptev Strait. On the way he saw the hulk of a ship cast up on shore with anchor and rigging still visible, the wreck of some trading vessel of the previous century. On September 2 he was off the Indigirka River. Here north winds brought down the ice and the ship, cut off from the land by shoals, was firmly beset. With the help of his geodesist Kindyakov, who was making a parallel survey by land, Laptev got his party ashore and wintered on the Indigirka. In the spring he directed surveys in the delta of the river, while Kindyakov, the pilot Shyerbinin and the capable seaman, Alexei Loshkin, explored the coast eastward toward the mouth of the Kolyma.

In the summer the ship was patched up and her liberation effected by cutting a canal through the ice. On shipboard Laptev passed the mouth of the Kolyma and a little beyond was halted by ice at the Great Baranov Rock. He wintered on the estuary at Nizhne Kolymsk. In the spring he built lodkas (boats) for a final attempt at eastward penetration by sea. It was unsuccessful and the project was finally given up. East Cape was still hun-

dreds of miles distant to the east, and the survey of the inter-
vening coast by land was dangerous owing to the hostility of the
Chukchi. Laptev turned his attention to the finding by dog team
of a route over the highlands to the upper waters of the Anadyr.
From there in 1742 he boated down to the Pacific and made his
way back to St. Petersburg.

" 'Peter's fledglings' " [the officers of the young Russian navy]
"had carried out with honour and glory the commands of Peter.
For a period of almost 200 years [until the 1910–15 voyage of the
Taymyr and the *Vaygach*] the observations and coastal charts
gathered and prepared by the members of the Great Northern
Expedition remained almost the only guide for seafarers in
North Siberian waters. The results of their work provided M. B.
Lomonosov with his materials when, in the years 1755–1763, he
produced a reasoned argument for the feasibility of traveling to
the Orient through the ice of the northern sea.... Unfortunately
only a small part of their observations have become the possession
of Science. For the most part copies only of the detachments'
journals survive and in some cases not even these remain."*

The stretch from the Baranov Rock, just east of the Kolyma,
to Bering Strait was left without official survey, and for nearly
a century geographers were vexed with the phantom of an isth-
mus sprouting from the North Siberian shore and arching over
to form a bridge to America. Incontrovertible proof that the
continents were divided was finally provided by Baron von
Wrangell in 1823.

* Ibid., p. 78.

VOYAGE TO AMERICA: CHIRIKOV ON
THE ALASKAN PANHANDLE

While "Peter's young men" were laying bare the Siberian north shore Bering was busied with the other branch of the projected discoveries, the construction of ships at Okhotsk and their dispatch on two missions, one to determine the relationship of Russian possessions on the Pacific with the island empire of Japan, the other a trans-Pacific voyage and a landing on the west shore of America. The general plan, matured by the end of 1732, provided among other things for a staff of scientists to share in the work of discovery. In February 1733, Martin Spanberg, who was to make the voyage to Japan, left with the advance party. Bering and Chirikov, with a host of ironworkers, shipbuilders and heavy equipment, set out later in the spring.

The transportation of an expedition of this size for more than the length of a continent was an enormous task, occupying several years and vastly exceeding the estimated cost. Bering, a plain sea captain, proved quite unfit to cope with the ill will of Siberian officials who could plead sheer inability to satisfy authorized requisitions.* Spanberg reached Okhotsk in 1735 and began the construction of shipyards. Bering, delayed by his responsibility for the northern surveys, did not arrive until 1737. Spanberg embarked for Japan in 1738; Bering's ships were ready for service in 1740. He planned to winter on the east shore of

---

* The difficulties of Bering are admirably set forth in Robert Murphy's *The Haunted Journey* (Doubleday & Co., 1961).

Kamchatka and from there set out on his trans-Pacific voyage. To further this plan he sent the mate Yelagin ahead in the *Gabriel* to make a survey of his intended harbor in Avacha Bay.

Either Bering had not been wholly sincere in the assurance he had given the Admiralty that the American shore must lie not far to the east of Kamchatka, or, as is equally probable, his honest optimism had wilted and his theories grown less convincing as the time for their application drew near. He now proposed to sail with two years' supply and spend a winter in America. This prudent policy was thwarted by the requirements of Spanberg and the loss of a cargo of biscuit upset in Okhotsk harbor. With his resources thus diminished Bering had no choice but to make the round trip to and from America in one season. In September 1740, he sailed with the *St. Peter* and *St. Paul,* the latter under Chirikov's command, rounded Cape Lopatka and anchored in Avacha Bay opposite the storehouse built by Yelagin. In honor of the ships the harbor was named Petropavlovsk.

The officers assigned to the ships were: on the *St. Peter,* Captain Commander Vitus Bering, commander of the expedition, Lieutenant Sven Waxell, Fleet-Master Sofron Khitrov, and Surveyor Plenisner; on the *St. Paul*, Captain Lieutenant Alexei Chirikov, Lieutenants Chikachev and Plautin, geographer Louis de Lisle de la Croyère, and mate Ivan Yelagin. In the course of the winter Bering enlisted another officer whom we owe a clear and graphic history of the expedition, a sympathetic understanding of Bering's character, and also, perhaps, the preservation of the *St. Peter*'s crew from a larger catastrophe than that which actually befell them. This was the German naturalist, Georg Wilhelm Steller.

Born in 1709 Steller had distinguished himself at the universities of Wittenberg and Halle in the fields of theology, medicine and botany. He migrated to St. Petersburg in search of advancement and was enrolled by G. F. Muller as his assistant in researches in the "three natural kingdoms" in eastern Siberia and Kamchatka. He was at Bolsheretsk when, in February 1741, he accepted Bering's invitation to the American expedition and was taken aboard the *St. Peter* with the rank of "adjunct."

It had not been the wish of the Admiralty College that Bering should sail east to make a blind landfall on the coast of America. Their instructions required him first to sail up the Siberian shore to East Cape, cross to Alaska and trace the American shore southwards, thus providing a broad and coherent picture of the relationship of the two continents. But, fearing obstruction by ice if he reached East Cape early in the season, Bering determined to reverse this course by sailing directly to America and then circling north, giving the ice in northern waters time to disperse. He had intended to winter at a conveniently southern latitude on the American shore and devote a leisurely two seasons to his circuit of the North Pacific, but, as we have seen, had been compelled to alter his purpose by the inadequate provisioning of his ships. The prudent course of deferring the voyage for another year in order to bring the ships up to the maximum of fitness in equipment and supply was ruled out by the years already wasted and the enormous costs incurred. The ships weighed anchor on May 29, 1741, the shadow of disaster already overhanging them. After much towing and warping against contrary winds they cleared Avacha Bay and set a course to the east on June 4.

The season was already dangerously advanced for ships which were to cross an ocean of undetermined breadth and be back at the home port by autumn. They were also hampered by a requirement which the Academy of Sciences had imposed and Louis de Lisle was there to enforce. Ships blown off course in the mid-Pacific had brought back word of land about 46° north latitude, called by various names and known to Bering as Juan de Gama Land. A clause in his orders required him to search for this land and verify the report. Had he taken the clockwise route, he would have approached the supposed land late on his homeward voyage with the option, in emergency, of bypassing it. By reversing his intended course he had made Gama Land (some five hundred miles south by latitude from Petropavlovsk) his first point of call—a vexing hindrance to an expedition already retarded and hurrying on a more important mission. On June 13 the ships were in the vicinity of 46° north latitude, 175° east longitude, with no land in sight. From the *St. Peter* Waxell hailed Chirikov directing him to give up the search and steer east by north. On the nineteenth the two ships parted company in fog and after some days of groping gave up hope of restoring communication. The overly conscientious Bering made another lunge to the south in search of Gama Land and so fell behind Chirikov who was speeding away on the appointed course.

The *St. Paul* inclined a little more to the north as she increased her easting, and at 1 A.M., July 15, raised land which soon assumed the shape of a prolonged mountainous coast, well treed on its lower slopes. She drew inshore toward the islets which lined the outer coast of Prince of Wales Island, with Cape Bartoleme in sight to the north. Finding no safe anchorage, Chirikov edged away to the north and on the sixteenth saw Cape Ommaney looming up through the fog. Leaving it behind he coasted up the shores of Baranov and Chicagov Islands, and past Cape Edgecombe. At noon on the seventeenth he was, by Golder's estimate, near the entrance to Lisianski Strait, which divides Yakobi Island from its larger neighbour Chicagov. Mountains were visible in the background stretching away in an endless series, northwest by north.

On the afternoon of the eighteenth the longboat was sent ashore to look for a safe anchorage and fill water casks. Fearing that he might be driven from his station by a squall, Chirikov studied the land features as the boat pulled away. Before him lay a rugged and forbidding shoreline, beneath him a rocky, irregular bottom where no anchor could grip. In storm, cloud and fog the *St. Paul* was driven seaward thirty miles. After two days of hard work against contrary winds Chirikov recovered his station and drew as close inshore as he dared, firing signal guns. On the twenty-fourth as there was still no trace of the missing boat crew or of native inhabitants, the small (and only other) boat was manned and ordered away. It was lost to view and never seen to reach land. By night the glare of a fire was seen, and the ship groped her way even closer among submerged rocks firing signal guns without effect. On the next day natives were seen in two canoes. They came within hail shouting, "agai, agai," and paddled off turning a deaf ear to the strangers who were pleading for a parley. Chirikov was now convinced that his men had been forcibly detained, if not murdered. He was powerless to rescue or avenge them. The ill-equipped *St. Paul* carried only the two boats of which she had been so mysteriously deprived. Chirikov hung out a lantern by night and lingered two days more on that fog bound and depressing shore—not solely for his comrades's sake. The missing boats were their only means for replenishing the water supply of which not more than forty-five casks remained. The acuteness of the emergency ruled out the debate and uncertainty which were to vex the cruise of the *St. Peter*. On July 27 the captain and his four most senior officers came to a unanimous resolution to make for their home port with all the speed possible. Water was rationed and orders given to catch rain water at every opportunity.

No trace has ever been found of the fifteen men whom Chirikov unwillingly abandoned to their fate; nor is there the vaguest native tradition of the capture or murder of the European strangers. The probability is that both boats were caught in the tide rip before reaching land. The spring tide at the entrance to Lisianski Strait flows at the rate of twelve knots. The Russians were singularly unfortunate in the place they chose for a landing.

By slanting their outward voyage toward the supposed Gama Land, Bering's ships had passed well to the south of the Aleutian Islands, which otherwise they must certainly have discovered.

But in his quest for an anchorage on the American shore Chirikov had made some 200 miles of northing, so that the southward arc of the long archipelago now lay between him and home. On August 1 he jogged sharply to the left to clear Kenai Peninsula, resuming a westerly course he came up against Kodiak Island and making no mistake this time he hauled off hundreds of miles down to 53° latitude and for the rest of August met with no other hindrance in his homeward voyage than contrary winds.

Late in the evening of September 9 the *St. Paul* suddenly found herself in shoal water. The murmur of surf was heard not far ahead. Chirikov gave orders to anchor for the night and with the coming of daylight the dissolving mist revealed that the ship was embayed with an island of the Adak group barely a quarter of a mile ahead. Men were visible on shore and in response to the Russians' hail seven of them put off, each in his own kayak, and drew near to the ship. Chirikov studied the natives with kindly interest. At first they seemed to be imploring the mercy of the powerful strangers who had come in so huge a boat. Then, comprehending that they were in some sort of distress, they paddled alongside and began to traffic, exchanging clothing, arrows and samples of mineral for the baubles which the Russians had provided for the express purpose of barter. In their extremity the latter tried hard to persuade the natives to bring off what water they could in their tiny kayaks; but the relief promised by such means was slight. The wind was shifting and Chirikov judged it prudent to make sail and work clear of the land without loss of time. Angling north to recover the fifty-third parallel they passed within twenty miles of Kiska Island without seeing it, sighted Buldir Island and veered off course to the south to clear the Semichi islets. On the night of September 21 they anchored again in shoal water and at daybreak found themselves on the east shore of Agattu Island with the "high snow-covered mountains" of Attu visible far to the northwest. Agattu was the last geographical obstacle. The *St. Paul* rounded the island to the south and after wrestling for some days with contrary winds steered an unimpeded course for Petropavlovsk.

But long before they cleared the reef-strewn Aleutians the crew of the *St. Paul* were menaced by a danger against which nautical skill provided no resource, scurvy. The log for September 16 records the death of the sailmaker and adds: "Captain Chirikov, Lieutenant Chikachev, and the members of the crew are very ill, owing to the lack of water and the long hard sea voyage. They are no longer able to work." On the twenty-sixth Lieutenant Plautin is also listed as desperately ill with scurvy. The men had scarcely the strength to mount to the deck and work the ship. Chirikov had been confined to his cabin from the twenty-first. The ship was navigated by the mate, Ivan Yelagin. "I gave him such help as I could, for thanks to God, my mind did not leave me. I worked out the courses from the log-book and told him what to do. [The mate was evidently a promoted seaman with no scientific training.] Yelagin has good judgment and if it had not been for him and the strength which God gave him, some great misfortune must have happened to the ship . . . I promoted him to the rank of fleet lieutenant, for he not only did the work of that office but ran the whole ship."

And he ran it so well and so inspired the will to live in his crippled and despairing crew that on October 8, sixteen days after passing Agattu, they raised the coast of Kamchatka. Pitiless headwinds held them off shore until the night of the tenth, when they cast anchor in Avacha Bay with one barrel of water only remaining. The two lieutenants had died at sea. De Lisle de la Croyère died only an hour after the berthing of the ship. But the *St. Paul*, as it proved, had come off luckily in comparison with her former consort. Of the twenty-two men she lost on the cruise only seven actually died of scurvy. The other fifteen had been lost with the boats on the American shore. Autumn and winter passed away without word of Bering or his ship, and in the spring of 1742 the *St. Paul* set out again on the unpromising mission of seeking her consort in the vast ocean to the east. With maimed ship and enfeebled crew it is to Chirikov's credit that he voyaged far enough to visit and fix Attu, the last of the Aleutians. No trace was found of the *St. Peter* there, and they returned to Petropavlovsk to find her still unreported. Chirikov gave up and set a course for Cape Lopatka and Okhotsk. A few days after his departure from Avacha Bay, Waxell and forty survivors of the *St. Peter* arrived there in the forty-foot hooker built from the wreck of their ship.

VOYAGE TO AMERICA: BERING'S LAST CRUISE

On losing Chirikov, Bering, uncertain in what direction he was to be found, steered southward, partly to convince himself beyond all doubt that Gama Land was not to be found. Thus when he resumed the appointed course, he was several days behind the *St. Paul*. From June 25 to July 13 he steered east by north, notwithstanding the remonstrances of Steller, who, observing in the water reed grass and seaweed of a type which occurs only on rocks and in shallow water, and noting also the presence of gulls and seals, insisted that they were traveling parallel to a continental shore not far to the north. He was almost literally right: the Aleutians lay not far below the horizon. If the seamen had heeded his advice they could, with the advantage of long days and fair weather, have learned the pattern of the elongated archipelago, along which they were to blunder so disastrously in the storm and darkness of the homeward voyage. But Steller had not the tact to make his advice palatable to professionals; the officers mockingly reminded him that he did not sit in "God's privy council"—a truth which Steller, with all his gifts, was prone to forget. He had no resource but to meditate bitterly on the opportunity that was being thrown away and on the stupidity of those who thought that a knowledge of navigation conferred mastery of all the sciences.

Though not unkindly disposed toward Steller, Bering was an unlearned seaman like the rest and could not always refrain from taking part in ridiculing of the intelligent but arrogant and humorless man of science. But as the weeks passed by with no land appearing ahead, Bering became alarmed at the ever-lengthening distance from home, and on July 13 did consent to haul up on a more northerly course. On the sixteenth land was sighted, on the seventeenth Mount St. Elias was plainly visible thrusting its snow-covered head above the level of the sea. As the ship drew nearer to land an endless chain of mountains took form, snowcapped, and with their lower slopes clothed with a beautiful forest extending to the very margin of a rocky and indented shoreline. The ship's advance was retarded by light variable headwinds: Bering slanted his course to north-north-west, and on the twentieth drew near to a high island with a single mountain, rocky, but covered with spruce. He rounded its southwestern end, the abrupt headland of Cape St. Elias, and—more fortunate than Chirikov—found a convenient anchorage low down on its western shore. Steller was much offended at the captain's lack of enterprise in not pushing his ship into the strait which separated the island, Kayak Island, from the mainland. (Modern survey finds this channel shallowing to a depth of four feet.)

The years of toil, frustration, and abuse which the command of the expedition had cost him had wearied Bering of the employment which he had eagerly solicited ten years before. In response to the congratulations of Steller and Plenisner, the tired old sailor grumbled that the voyage had been much longer than he had anticipated, the ship was poorly supplied and far from her home port, and that trade winds (winds blowing constantly from the same quarter) might retard their return and reduce them to the direst extremity. Now that he had fulfilled the principal clause in his orders by reaching America the anxious old sailor had little thought but to get ship and crew back safely with the least possible delay.

In the meantime, says Steller, the officers had been openly and noisily debating the proper course to pursue, instead of presenting their views to the captain in a decent and orderly fashion.

Bering terminated the debate by directing Khitrov to survey the island in the large boat with an armed crew, while Waxell took charge of the yawl and replenished the water casks from the nearest source available. The officers contemptuously rejected Steller's demand to be taken ashore, asking him what *he* hoped to accomplish and tried to frighten him with "dreadful tales of murder." Steller in turn demanded to know if they had traversed hundreds of leagues of ocean merely to carry American water to Asia, declared that he was "never so womanish" as to shirk danger in fulfillment of his duty, and appealed to Bering for permission to land. The captain had weakly countenanced the ridicule to which the scientist had been subjected, but he had none of his subordinates' petty meanness: he granted leave to go ashore with the caution that Steller was not to stray too far and delay the departure of the ship. Steller seated himself beside Waxell in the yawl with the bitter consciousness that no one aboard understood or cared for his work as scientist, and that he had been enlisted in merely formal compliance with orders and to make up for the deficiencies of Assistant Surgeon Betge, whom no one trusted.

The Kayak Island where Steller made his landing is a narrow ridge heaved up out of the sea, high, well wooded, and extending some twenty miles from its southwestern extremity, Cape St. Elias, 59° 55′ north latitude, 144′ 30° west longitude, to the continental shore at Cape Suckling, from which it is separated by a shallow strait. The point where the German scientist disembarked was identified in 1922 by Leonard Stejneger as lying on the west shore seven miles above Cape St. Elias. As soon as he was ashore Steller set out up the beach, accompanied only by his Cossack servant, and soon found proof that the island was inhabited. He came upon a hollow log, evidently used as a cooking utensil, and reindeer bones. A fire, still burning feebly, gave warning that natives, almost certainly strangers to white men and easily frightened into ferocity, might be lurking in the woods nearby. Undaunted he continued up the shore, observed trees clumsily felled with stone axes, and with some misgivings followed a path which led away from the beach into the deep twilight of the primeval forest. There he found a cellar, covered with poles and cut grass and containing bark utensils, smoked fish, arrows, thongs of twisted seaweed, and sweet grass, evidently gathered for distilling purposes.

In this eerie situation, in which the bravest of men would find loneliness appalling, the bold Steller sent his Cossack back to the landingplace with samples of what he had found, and alone tramped four miles up the beach to a point where an abrupt cliff cut off further progress along the shore. A roundabout scramble over rock and through thicket brought him to the summit of the headland where he saw that the wall of cliff extended so far as to make descent to the beach ahead impossible. He attempted to cross the island to its east shore, but was deterred by the denseness of the forest, where, without a compass, there was danger of losing his way. He went back to the landingplace where the yawl was just putting off with water casks and sent a message to Bering requesting that he might have the use of the boat to sail around the island and collect plants. In the meantime he kindled a fire and "tested out the excellent water for tea." An hour later Waxell returned in the yawl bearing a curt message that Steller had better come aboard without delay unless he wished to be left behind. He had been ashore for ten hours. A little later Khitrov returned with the large boat after rounding the island and preparing a rough but accurate chart of it and the adjacent continental shore.

Early on the morning of the twenty-first, Bering, with no previous consultation, gave orders to hoist anchor and put out to sea. Steller was bitterly disappointed at this craven decision which destroyed his hopes of noteworthy scientific discovery. Waxell angrily protested that twenty water casks remained unfilled. Bering refused to alter his decision: he would not miss the fair wind then blowing, and declared that in view of the lateness of the season they would not make a survey towards East Cape but steer straight for home. It was soon discovered that the direct route for Kamchatka lay along the Alaskan shore as it curved away to the west and southwest. Apparently without the knowledge of the captain who kept much to his cabin, the officers of the *St. Peter* kept the ship close to the land, losing much time in shoal water, and sometimes compelled to anchor in darkness or fog. They were reluctant to leave America without further discovery, for in the Russian service junior officers were apt to share in the punishment of a delinquent captain. On August 11, finding themselves in danger of being embayed, all agreed to give no more thought to exploration, but to make for Avacha Bay with all possible speed.

The conflicting wishes of the captain and his officers had been reconciled too late for the safety of the expedition. Bering had ordered a hurried departure from Kayak Island in the hope that fair wind would carry him far on the homeward journey. Waxell and Khitrov had squandered this advantage by three weeks of slow and cautious navigation close under the land. And now, when all were agreed on the necessity of a speedy return, the wind turned foul and so baffled them for two weeks that their dwindling water supply was plainly insufficient. So they turned north and back to the land and anchored amid a cluster of islets, with the shore of the Alaskan Peninsula dimly visible on the northern horizon.

Khitrov, detested by Steller, but evidently a man of enterprise, obtained Bering's surly consent to make a survey of the islands in the yawl—if he was willing to take the risk of being left behind. Other boats were detailed to fill the water casks. Steller quickly found fresh springs, but on returning to the beach was enraged to find the men filling the casks from puddles situated conveniently near the beach, but brackish from ocean spray. Going aboard, Steller found Bering too sick and indifferent to listen to his complaints; his request for a detail of men to gather the antiscorbutic herbs, already badly needed, was "spurned and coarsely contradicted." As a storm was rising, the sick were brought on board, but Khitrov failed to return. At twilight the sight of a large fire on one of the islands assured his comrades of his safety. The anchorage proved to be sheltered enough to relieve Waxell of the need for moving the ship and, in the wild weather then prevailing, being driven out to sea for good. On the next day the large boat rescued Khitrov and his crew. The yawl, unseaworthy in the wind-swept channel, was left behind.

The crisis did not end with the rescue of Khitrov. The wind blew strong from the southwest; tacking to windward in a narrow channel was out of the question, and as Waxell was not willing to seek another outlet by going deeper among the islands, he had no choice but to remain where he was. On the night of September 3 a fearful gale arose. Topmasts and yards were struck and with two anchors out the heaving ship clung to her station. On the sixth Waxell found a side exit to the southeast between two islets and at last recovered the freedom of the open sea.

The *St. Peter* had made her American landfall at a higher latitude and consequently at a more westerly point than the *St. Paul*, and as the two ships circled north and west following the curve of the Alaska shore the commodore vessel was well in the lead. But while she was berthed at the Shumagin Islands, the *St. Paul* passed her far out to sea, and on September 8 when Waxell again got his ship back on a westerly course Chirikov was 500 miles ahead. The *St. Peter* had wasted the best weather vouchsafed to the two ships on the way home. Nevertheless for two weeks she made uninterrupted progress, though latterly hampered by west winds and increasing sickness among the crew. On September 24 in 175° east longitude, forced by a southwest wind to steer north, she raised high land ahead, the Atka and Adak islands and numerous smaller islets. As the wind was foul for going west, the luckless ship was obliged to wear and scud away to the southeast to avoid being embayed. For more than two weeks she was driven away from her home port by furious gales. It is a tribute to the work done in the improvised dockyards at Okhotsk that she remained tight and seaworthy.

Had she been manned by a full complement of healthy men she might have fought into the wind and diminished the westing lost, but by September 30 half her crew were sick or incapable of exertion and the rest "quite crazed and maddened from the terrifying motion of the sea and the ship." Consequently on October 11 when the ship was again able to eat into the wind, she had been driven back on her course a full seven degrees, and not until October 18 did she recover the westing of September 24.

The crisis was now acute. On October 24 Steller states: "Misery and death suddenly got the upper hand on our ship to such an extent that not only did the sick die off, but those who, according to their assertions, were well, on being relieved of their posts dropped dead from exhaustion. The small allowance of water, the lack of biscuits and brandy, the cold, dampness, nakedness, vermin, fright and terror were not the least important causes."

The *St. Peter* had dropped below 51° north latitude to get clear of the island chain. On the twenty-fifth while on a northerly course to recover the latitude of Avacha (53° north) they saw land ahead, Kiska Island. Passing to the left on the twenty-eighth they sounded in fourteen fathoms and saw land a mile ahead. Khitrov's advice to put in for water was rejected on the reasonable ground that if the anchor was dropped the "ten feeble persons" who still kept the ship going would not have the strength to raise it. They rounded the obstacle and still on a northerly course cut through the tail of the Aleutians and passed north of Attu Island without seeing it. Course was altered to the west and the ship ran on through rain and wind, the latter fortunately steady as the crew could not muster the strength to shift the sails. On November 4 snow-covered peaks appeared above the western horizon and caused universal joy. No one doubted that they were approaching Avacha Bay. The officers were congratulating themselves on the miraculous precision of their reckoning when a rift in the clouds afforded them a sight showing their latitude to be 55 north, not the 53 of Avacha. Steller listened with ill-natured relish to the labored explanations of the disconcerted navigators. Lacking the sea room to coast southwards they tacked back and forth during the night in a violent storm.

Daylight showed them close in toward the land and opposite a bay protected by shelving rocks which extended half a mile into the sea. Khitrov assembled a council which Bering and Waxell attended, though both were desperately ill. Khitrov wished to winter in the haven ahead: half the ship's company were unfit for the lightest duty, the foremast was the one trustworthy spar standing, and only six casks of water remained. Bering, possibly convinced that he had not long to live, was for sailing on. The demoted Ovtsin, whom he had made his adjutant, supported him in urging the continuance of the voyage. Thereupon, says Steller, Khitrov bade him hold his tongue and drove him from the council. He then appealed to Steller, who replied that his advice had invariably been received with contempt, and that he now preferred to keep silence. He was then asked to certify in writing the condition of officers and men, and this he consented to do. His discouraging diagnosis, coupled with Khitrov's confident assertion that they had reached the mainland, decided the council to put an end to the voyage and berth the ship in the bay ahead. Even supposing that Steller's detestation of Khitrov had not distorted his version of these transactions, the Fleet Master's arrogant self-assertion claims indulgence. He was the senior officer capable of duty on deck; on him fell the task of carrying out whatever decision was made, and he knew better than Bering or Waxell the feebleness of the men and the wretched condition of masts and tackle. In asserting positively that they had reached the continent he may have assumed a confidence he did not feel to procure the adoption of the best course offered.

So they edged inshore and when night came on cast anchor, not daring to enter the rock-strewn and uncharted bay in darkness. When wind and surf arose the cable parted, and a second anchor was lost in trying to bring the ship to. In this emergency, asserts Steller, Khitrov lost his head: Ovtsin and the boatswain took charge, and bade the crew spare their anchors and let the ship drift. She thus drove across the bar and found anchorage in the still waters within. The seafarers were even luckier than they knew; Waxell states that nowhere else on the east shore of Bering Island could they have found a berth for the ship backed by a level beach for those who manned her.

On November 6 Waxell exerted himself to rise and go ashore with Steller. They landed on a broad expanse of sand backed by hills which mounted up to terminate in a ragged mountain ridge. They soon found what they most urgently required, a freshwater river; for the rest, the aspect of the land was forbidding. Apart from the arctic willow it bore neither tree nor shrub; driftwood piled along the beach provided the means of fuel and shelter but it was buried in snow and hard to gather by sick and exhausted men. Wild life however was plentiful along the beach: blue foxes literally swarmed on the sandy flats; ptarmigan were plentiful; and the sea otter were seen in numbers, all so tame that Steller quickly inferred that they stood not on the continent, but on an island unvisited by predatory man. On the eighth Steller and Plenisner supervised the landing of some of the sick. On being brought out, several were killed outright by breathing the raw winter air. Instead of casting the corpses into the shallow water of the bay the two Germans and their helpers, with a humanity most creditable in such an emergency, carried them ashore for burial. Bering was landed on the tenth and housed in a tent where he passed a miserable night in gale and heavy drift.

The excellent Steller, who, with all his defects of temper had not a grain of vanity in his composition, makes no boast, but it is fairly evident that, in this crisis, with half a shipload of sick men being brought ashore and laid on the beach by comrades almost as infirm as they, he was the principal agent in restoring health and morale. He had strength for the task for he had kept himself, Plenisner and their Cossack servant Petrusha in fair health by use of herbs gathered on the Shumagin Islands. Khitrov was void of ideas and spoke of wintering on the ship with Waxell. He had some reason, for furious winds were making the shelter of hut or tent insecure. Observing the numerous foxholes on the river bank Steller proposed burrowing pits in the sand and roofing them with canvas and driftwood. Soon the more capable men began the systematic excavation of huts for themselves and their disabled comrades. On November 12 the last of these were ferried ashore. Several died, some on reaching the deck, some in the boat, and others on the beach. Foxes, tame and fearless, mutilated the dead and even gnawed at the bodies of those who lay strewn still alive on the sand. Consigning these unfortunates to the care of Assistant Surgeon Betge, Steller busied himself in promoting the hunt for fresh meat which the tameness of all animal life made easily procurable. The priceless skins of seal and otter were cast aside to be torn to rags by the foxes while the carcasses were dragged to camp and distributed among huts and tents. In the intervals of hunting the surgeon prepared soups for Bering and the worst afflicted of the scurvy patients.

From his deathbed the worthy old captain still took thought for the men in his charge. On the twenty-fourth he gave orders to bring the ship in from the exposed sea where she lay at anchor. Five men placed Waxell in a boat and rowed out to execute the order. But they had not the strength to hoist the anchor; they dared not cut the cable with the surf rolling on a reef to leeward, and so were obliged to leave her where she was despite the danger of her being torn from her anchorage and blown out to sea. "By a stroke of luck" when her cable did part four days later she was driven inshore and piled up on the very sandbar where it had been intended to beach her.

Steller, teaching by example since his advice was disregarded, organized the inmates of his hut, himself, Plenisner, Betge, and five seamen, into a community pledged to work for one another and for the crew at large. To promote ease and harmony he directed that titles of respect should be dispensed with, and instead of the patronizing "Petrusha," solemnly addressed his servant as "Peter Maximovich." When the camp was finally organized the officers and their attendants occupied four huts and the remainder of the crew a larger barrack. Its dimensions are not given, but it must have been small and its inmates severely cramped to be securely roofed in a climate of gale and heavy snow.

The establishment of some degree of order and comfort came too late for some: deaths continued to occur. On December 8 the Captain Commander Vitus Bering died. Forgetting former peevishness Steller took kindly notice of him, "a devout Christian—kind, just, universally liked by high and low" but "not born to quick decision and swift action, faithful, dispassionate, circumspect and deliberate." By then the plague of scurvy which had so ravaged the crew had been effectively checked. Sixteen men had died in November after the landfall; five died in December and two in January, the last on the eighth of the month. Forty-five remained of the seventy-six or seventy-seven who had embarked on the cruise, and all of these were to come safely to Kamchatka.

Winter passed without hardship or serious disorder to the crew. Temperatures on the island proved mild. Roofed-in dugouts gave the needed shelter from gale and snow, and circumstance delivered the men from idleness in these cramped quarters. The hunting of the sea otter kept many occupied until March when the reappearance of the "sea bear" (fur seal) and the sea lion gave the hunters a wider choice of game. Units were also formed to bring in the driftwood which lay up and down the beach. Their journeys of necessity grew longer until late March when the thaw uncovered tree trunks that had been overlooked or were too solidly embedded to be dragged out. Mist, snow and sudden gales were a constant threat to those who ventured too far from the camp. Early in April a party including the only surviving carpenter was cut off in a blizzard and absent for days. This caused a small panic among the men, for by then they were convinced that they *were* on an island, that their ship was unserviceable, and that unless a boat could be put together from her timbers they would never see their homes again.

Bering Island, where the *St. Peter* had been cast away, is the largest of the Commander group where the Aleutian ridge, submerged all the way from Attu, rises, as it were, for its last gasp a hundred miles from the Kamchatka shore. The ship had come to rest on the east side of an island some hundred miles in length; and as travel along the beach under the cliffs was difficult and across the mountains next to impossible, proof positive that her crew were cut off from the continent was only obtained at the end of March when the boatswain in a journey of some weeks took a party to the north end of the island and found ocean to the west with no land in sight. He confirmed what the rest had already come to believe.

In January surveyors reported that the ship was seriously damaged and embedded in sand beyond the power of her crew to refloat her. Yet such was the dread of government displeasure that some raised the question of whether Her Imperial Majesty's ship could lawfully be taken apart and urged that relaunching be attempted. The majority of officers and men voted with Waxell that the wreck was to be broken up and a smaller craft built from her timbers.

Both ship's carpenter and carpenter's mate had died on the voyage; the castaways had to rely on the practical skill of Starodubtsov, the Cossack who had learned something of shipbuilding in the dockyard at Okhotsk, and who promised, given the dimensions, to build a ship "so solid that, with God's help, we should be able to put to sea in her without risk." This intelligent and enterprising frontiersman was subsequently on Waxell's recommendation made "a Siberian nobleman" by the chancellery of Yeniseisk.

In April Waxell mustered the crew and detailed the fittest for continuous work on the ship, leaving to the rest the work of maintenance and the obtaining of meat. The latter proved no easy task; despite Steller's protests otter and seal had been so wantonly massacred that they were now far to seek and not easily caught when found. Eventually a device was hit upon for catching the four-ton sea cow (*manati*); when they came in with the high tide to feed on the sea grass which was their food, boat crews hooked these unhappy creatures with a barb attached to a hawser secured on land and hauled them in with a winch. By this means they were able to nourish themselves and their comrades and to add a little cow meat to the flour they had saved for the voyage.

On August 10 the ship was launched, 40′ in length, 30′ in breadth and 6½′ in depth, a one-masted hooker with mainyard and mainsail, flying topsail, foresail and jib. She proved to be a good sea boat, and ten years later was still on the Okhotsk-Bolsheretsk run. She also bore the name *St. Peter*.

They put out to sea on August 13, sighted Kamchatka on the seventeenth, and on the twenty-seventh cast anchor in Petropavlovsk harbor. The homecoming for months so ardently prayed for was something of an anticlimax. Chirikov, giving up the *St. Peter* for lost, had departed for Okhotsk a few days before, carrying with him, naturally enough, all the belongings of the *St. Peter*'s crew to preserve them from plunder by the Kamchadals. The deprived wayfarers spent a comfortless winter at Petropavlovsk before being carried to Okhotsk and disbanded to straggle homewards or seek other employment on the Pacific shore.

The formalities attending the close of the Great Northern expedition were as protracted as had been its preliminary operations. It was not officially terminated until 1748. In that year Chirikov died. The reinstated Ovtsin and other officers who had shared in the service lived to attain high rank in the Russian Navy. But the principal historian of the *St. Peter*'s voyage, who had made it a success in the field of science and had done so much for the health and preservation of her crew, came to an early and pitiful end. No one could have blamed Steller if, after the hardships he had suffered in uncongenial company, he had applied for the immediate discharge which his German friends on the St. Petersburg Academy of Sciences could easily have procured him. Nothing was farther from his thoughts. Greed for fame, or more probably the sheer passion for scientific discovery drove him on to years more of toil. He spent the winter of 1742–1743 working out of Bolsheretsk. In the next summer he crossed over to continue his work on the Kuril Islands. In the autumn he came back to Kamchatka, built a boat, hired a crew at his own cost and wintered with them on Bering Island. Then with a constitution broken down by labor and the drunken excesses he had used to alleviate it, he began the journey back to European Russia. He was twice stopped on the road and once turned back hundreds of miles to answer charges of having obstructed officials in their dealings with the Kamchadals. Eventually cleared, he had passed through Tobolsk for the third time, when in the winter twilight his driver pulled up at a wayside inn and went in to warm himself, leaving Steller in the sleigh, heavily wrapped and his senses dulled with liquor. The driver came back to find his passenger chilled and in a deep coma. Steller was hurried to Tyumen and there he died. He was thirty-seven years of age, in the prime of his working life. Carlyle's epitaph on the French revolutionary Bailly may well be applied to Steller: "Crueller end had no mortal."

The Russian explorers of the period 1725–1742 introduced a new epoch in global history. The concern of earlier discoverers had been chiefly with commerce, precious metals or piracy. The Emperor Peter (prompted admittedly by foreign savants) gave his country priority in geographical enterprise aiming primarily not at trade or plunder, but at the extension of knowledge. Steller was as worthy a representative as either Peter or Bering of the era which all three helped to introduce. He was the prototype of the great school of surgeon-scientists, John Richardson, Elisha Kane, and Edward Adrian Wilson and others, who in the far-reaching voyages of the late eighteenth and nineteenth centuries, besides faithfully discharging their appointed duties, supplied leadership in times of crisis, strengthened sagging morale, and enriched the literature of travel with their memoirs. Various circumstances, especially the fact that he appeared at the very dawn of the scientific revolution, have deprived the luckless German of due recognition. Had he lived half a century later he might have enjoyed worldwide applause and been accorded by posterity the highest rank in the school which he had founded.

THE NORTHEAST PASSAGE MADE:
COOK, WRANGELL, NORDENSKIOLD

During the generation following Bering's last voyage, the Russian traders who had overrun Siberia began to ooze their way in less resolute fashion along the Aleutian chain to the lower Alaskan mainland. For a period trading stations were set up on the islands only, to hinder the generally hostile natives from mustering against them in greater force than the intruders could cope with. Thus restricted the traders could do little for the advance of geographical knowledge: accurate delineation of the opposing shores of Asia and Alaska was left to the agent of a foreign power.

In 1778 Captain James Cook came up the American west shore with H.M. ships *Resolution* and *Discovery*, and crews including several men of distinction. Among the officers were Lieutenant James Burney, the future naval historian, and sailing master William Bligh. The ratings included the seaman Joseph Billings and that compulsive wanderer, the Connecticut-born John Ledyard. Alaskan geography was not Cook's prime concern: his mission was to traverse the Northwest Passage of America from its western outlet. But as it was thought that Alaska might be a large island divided from America by the fabled Strait of Anian, Cook kept close to the Alaskan shore and tested Prince William Sound and Cook Inlet, before crossing the Aleutians by Unalga Pass and setting a course for Bering Strait. He rounded Cape Prince of Wales, which he judged to be the westernmost point of all America, and steered up the northwest shore of Alaska until, 200 miles past Cape Lisburne he was halted by a barrier of polar ice pack pressing close to Icy Cape. Thwarted in his appointed duty but scorning to turn back (to Cook "inaction was death," one of his officers declared) the indefatigable explorer coasted west along the pack edge, across the Chukchi Sea and sighted land "in two hills like islands." This was Cape North, now Cape Schmidta, where the Newfoundlander Bob Bartlett was to make his landing after the wreck of the *Karluk*. Hindered by the wall of ice from rounding Cape North to the west, Cook again put about and traced over 200 miles of Siberian shore to arrive at the narrows of Bering Strait off East Cape, "a steep rocky clift next the Sea and off the very point are some rocks like spires." Cook concurred with Bering that here was "the Eastern Promontory of Asia." In a few weeks the master craftsman had converted the vague and disconnected observations of his predecessors into a coherent pattern, showing the relative positions of Asia and Alaska with a precision which we have barely improved on.*

Two hundred miles of Siberian shoreline extending from the Great Baranov Rock to North Cape remained unexplored. In 1785 the Empress Catherine commissioned the Englishman Joseph Billings, who had been Cook's Assistant Astronomer, to trace the coast from the Kolyma River to East Cape by water, if possible. He was also authorized to visit the Aleutians, make discoveries and report on the conduct of the Alaskan fur traders.

---

* Cook went south to winter at the Sandwich Islands, and there was killed in an affray with the natives in February, 1779. His successor in the command of the expedition, Captain Charles Clerke, though dying of consumption, took his ships back into the fog and ice of the northern sea without improving on Cook's performance on either the American or the Asiatic side of the Chukchi Sea. He died and was buried at Petropavlovsk. During one of his visits to the Siberian port Clerke had had the monument to the French geographer, Louis De Lisle repaired. In 1786 the discoverer La Pérouse, who had formerly treated the captive Samuel Hearne with such kindness, repaid Clerke's courtesy by redecorating the grave of the English captain. La Pérouse died himself a few months later when, in storm and darkness, his ships were dashed to pieces on the rocks off Vanikoro Island. The North Pacific was a most unlucky region for those who tried to probe its secrets.

Billings sailed from the Kolyma on 24 June 1787 with two boats and the Russian Sarichev as his second-in-command. He fared no better than Dmitri Laptev and was stopped by ice close to Chaunskaya Bay. Billings sought the advice of his officers on the feasibility of renewing the attempt: one of his councillors quoted the voyage of Dezhnev, whereupon Sarichev observed that in that instance "Nature had altered her usual practice"—or, as a non-Marxist would now express it, Dezhnev could only have got through by Divine intervention. The majority evidently sympathized with this view, and it was resolved that the voyage was best attempted from the opposite quarter, by traveling from East Cape to the Kolyma. Billings therefore journeyed to Yakutsk, and there met his former shipmate, John Ledyard, and would have enlisted him in the expedition had not the American been arrested by Imperial order as a spy. Billings made no attempt to procure his release on the reasonable ground that the Yakutsk officials were allowed no discretion in carrying out orders from the capital. Burney, who knew both men and evidently thought poorly of Billings, observed that the Empress would have shown better judgment by jailing Billings and transferring his commission to Ledyard.

In June, 1790, Billings sailed from Okhotsk to the Aleutians. This part of the voyage was unfruitful in geographic discovery and served no purpose other than to expose the cruelties practiced by Russian hunters on the native population. Billings returned to the continent, parted from his ship in St. Lawrence Bay, just south of East Cape, and with a mixed party of Europeans and Chukchis made the journey by *baidar* (native boat) and on foot to the Kolyma where he arrived in February. Unfortunately the route dictated by his Chukchi guides afforded him only occasional glimpses of the coast he was appointed to survey; his journey was useless, and Burney's condition that the separation of the continent could be absolutely proved only by "the ascertainment of a continuity of sea clear round the coast of Asia from the Kolyma to Bering's Strait" was still unfulfilled.

Billings's contemporary, Burney, and Viese, in modern times, have dealt perhaps too harshly with the pathetic futility of his prolonged and costly expedition. His real error lay more in accepting the Empress's commission than in his manner of discharging it. He was a deep-sea sailor (without Bering's long experience in the Russian service); he was most unwise in undertaking a duty in which any accident might deprive him of his ships and make him dependent on native guides and boatmen, whom any literate Russian trader, with his knowledge of the people and their language, was far fitter to direct than he. Billings's exposure of the "abject slavery" to which the Aleuts of Unalaska were subjected should be scored to his credit; it may have contributed to the milder conditions of servitude noted by Vancouver on his visit to an Alaskan trading post. And Wrangell pays him the singular honor of naming him along with Cook as the only two visitors to the North Siberian shore whose observations could be relied on.

It had been the Russia under Peter the Great which inaugurated the many voyages undertaken in the eighteenth century on behalf of science. It was Russia again which set the example of renewed discovery after the long moratorium of the Napoleonic wars. The seaborne expeditions of Kotzebue and Bellinghausen of the Russian Navy were launched, and in 1820 Lieutenants Anjou and von Wrangell were sent to carry on the work of discovery into the seas beyond the Lena and Kolyma Rivers.

From the earliest period of the Cossack invasion the Russians had been aware of distant land, beyond the Svyatoy Nos (Holy Cape). In 1712 a party under the Cossack Vagin, sent by the governor of Yakutsk, crossed Laptev Strait and spent some time on the Greater Lyakhov Island. Vagin can hardly have been an informed man of science himself, but he must have driven his men hard in the pursuit of knowledge, for, rather than repeat the journey in the next year, they murdered him, his son, and his chief lieutenant. For this the two ringleaders were executed "before a great concourse of people," their accomplices were knouted without mercy and banished to the then miserable out-post of Okhotsk. After this discouraging start further discovery was deferred until 1773. Then Ivan Lyakhov revisited the islands which bear his name, and in the next thirty years others, among whom Hedenstrom and Sannikov were prominent, extended the work and crossed Sannikov Strait to the larger New Siberian group beyond. Lieutenant Anjou's task was to prepare an accurate survey of these extensive new lands. His line of approach was by way of the Lena River. Wrangell's mission was to make his base at Nizhne Kolymsk near the mouth of the Kolyma, explore the unknown shoreline from the farthest point reached by Dmitri Laptev to Cook's Cape North, and to journey seaward in search of the Bolshaya Zemlya (Great Land), for two centuries an imposing fantasy which really existed only in the modest dimensions of Wrangel* Island. Wrangell's expedition is of particular interest, not from any novel or striking incident, but because he was a well-informed, sensible and benevolent observer, whose journal survived the neglect and procrastination of Russian officialdom to be published first in German and then in English.

The two officers with their respective parties separated at Yakutsk, Anjou taking the easy way by water down the Lena, and Wrangell setting out on September 12, 1820, by post-horse and pack train overland to the Kolyma. At a station on the Indigirka River he met the aged Father Michel, who, in a period of sixty years had baptised 15,000 natives of various tribes, and had, Wrangell added, really taught Christian truth and righteous living to his scattered and primitive flock. At eighty-seven the good father still traveled abroad to baptise infants, and supported himself by hunting and gardening. From posting-station to posting-station Wrangell crossed to Sredne Kolymsk on the upper Kolyma and descended the left bank of the river to tidewater at Nizhne Kolymsk. He reached his intended base on November 2, 1820, just eight months before Lieutenant John Frank-

* The spelling of Wrangel conforms to modern geographic practice.

lin boated down the Coppermine River to begin the mapping of the totally unknown polar seaboard of America.

Nizhne Kolymsk was a fort (*ostrog*), a wooden palisade with four towers, sheltering the administrative building and the storehouses. Outside stood the church and forty-two dwellings. In summer mosquitoes were abundant and vegetation scanty, with only reedy grass and willow growing on the southern slopes. Up the river at Sredne Kolymsk larch, polar and low-growing cedar flourished, and radishes could be grown in sheltered valleys. Animal life—reindeer, bear, fox, sable, squirrel, wolves, swan, geese and ducks—was plentiful. The natives lived on fish, wildfowl and reindeer. The male population of the Kolyma area totaled 2,498 males, 325 of them Russians. Wrangell later found the shoreline from the Kolyma to Shelagskiy Nos mossy and quite uninhabited: beyond the Nos native inhabitants were only infrequently encountered.

Wrangell's winter was spent in the task, familiar to men in his situation, of accumulating dogs, drivers and supplies. The local commissioner did not prove as obliging as his remote superiors (who were in the happy position of not being called upon personally to honor the pledges they gave) had promised, and Wrangell, who was evidently neither harsh nor arrogant, consented to reduce his demands and curtail his intended journeys. He later found out that his full demands could have been met. One can imagine, however, that a levy of fifty sledges, 600 dogs and a quantity of supplies might have caused more inconvenience in a small settlement than the commissioner was willing to inflict.

In December a gale from the north drove the sea up the river, bursting the ice and sweeping away the fishing nets. The natives took this loss with philosophy, asserting that such a phenomenon meant a larger spring catch, a prediction which, says Wrangell, was later fulfilled. From 19 February to 14 March 1821 Wrangell was away on a preliminary journey past Chaunskaya Bay to a point forty miles east of Shelagskiy Nos.

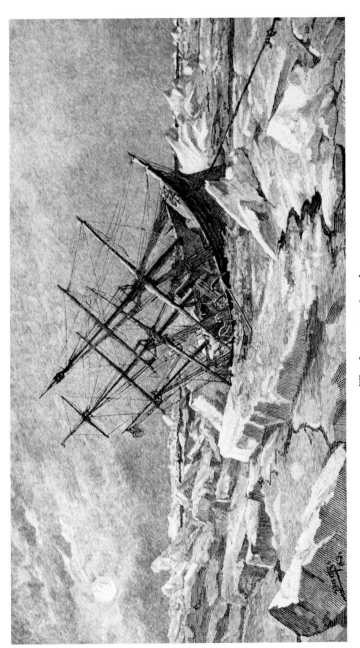

The *Jeannette* nipped

De Long's last camp

Fiala's *America* in the pack

A summer camp on Franz Josef Land

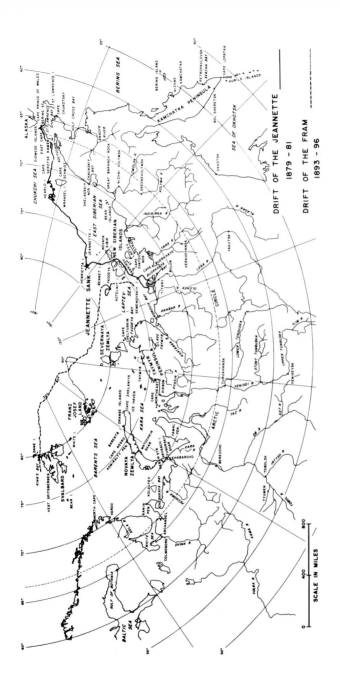

DRIFT OF THE JEANNETTE
1879 – 81

DRIFT OF THE FRAM
1893 – 96

SCALE IN MILES
0   400   800

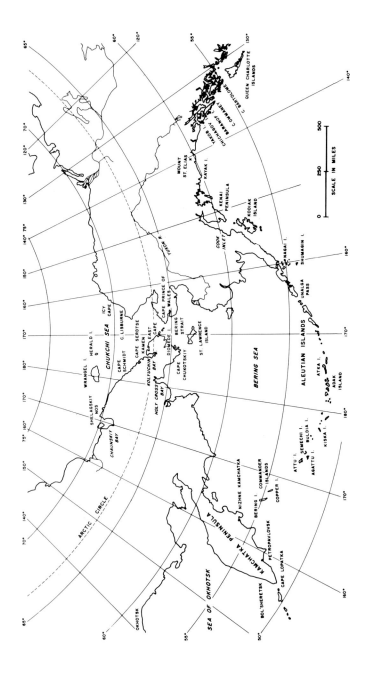

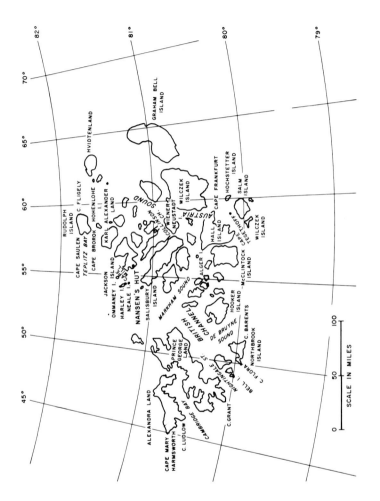

RUDOLPH
ISLAND

CAPE SAULEN
TEPLITZ BAY
CAPE BROROK
HOHENLOHE
I.

C FLIGELY
HVIDTENLAND

JACKSON
OMMANEY I. ISLAND
HARLEY I.
NEALE I.
NANSEN'S HUT
SALISBURY
ISLAND

KARL ALEXANDER
LAND

JDLING SOUND
RICHARDSON SOUND
WIENER
NEUSTADT
WILCZEK
ISLAND

GRAHAM BELL
ISLAND

CAPE FRANKFURT
HOCHSTETTER
ISLAND

AUSTRIA
SALM
ISLAND

HALL
ISLAND
CAPE
TEGETHOFF
WILCZEK
ISLAND

MARKHAM SOUND

ALGER I.
McCLINTOCK
ISLAND

PRINCE
GEORGE
LAND

BRITISH CHANNEL

HOOKER
ISLAND

DE BRUYNE
SOUND

C. BARENTS

ST NIGHTINGALE I.

BELL I.
C. FLORA
NORTHBROOK
ISLAND

C.GRANT

ALEXANDRA LAND

CAPE MARY
HARMSWORTH
C. LUDLOW BAY
CAMBRIDGE BAY

0   50   100
SCALE IN MILES

82°

81°

80°

79°

70°

65°

60°

55°

50°

45°

Franz Josef Land

There he made contact with the nomadic Chukchi, won their confidence and received assurance of friendly treatment on his coastal journeys. He found these roving people in possession of American goods obtained from Yankee traders on the Pacific shore. On March 26, Wrangell with twenty-two sledges set out northward from the Kolyma. Less than 200 miles out he was stopped by thin ice, and after a survey of the Bear (Medvezhyni) Islands went back to the Kolyma. A similar effort in March, 1822, made from the Great Baranov Rock was equally fruitless. With his officers, the midshipman Matushkin, mate Kosmin and surgeon Kyber, Wrangell spent the summers in survey and other scientific work east and west along the coast and up the valleys of the Kolyma and its eastern tributaries.

His great effort was reserved for late February–May, 1823. Matushkin and Kyber were detailed to complete the coastal survey east to Cook's Cape North. Wrangell himself, strengthened by the assurance of the natives that on a clear day snow-capped mountains could be seen seaward from the cliffs east of Chaunskaya Bay, rounded Shelagskiy Nos and struck out seaward for the third time. His advance was so retarded by a maze of hummocks (ridges of ice) that he ordered some of his sledges back to land and went on with the most active crews. (His track was parallel to and considerably to the west of that taken by Captain Bob Bartlett in 1914, when he made his escape from Wrangel Island to North Cape.) Like Bartlett he found the Chukchi Sea treacherous even in March. His route was straddled by thinly frozen and barely passable leads, and after one panic sprint over a broad expanse of "yielding ice" the dogs were so exhausted by fear that they sank down powerless on reaching firm ice. High winds and volumes of "frost smoke" (vapor given off from open water) gave promise of more trouble ahead. Soon, beyond the white of the ice field, a dark line appeared on the northern horizon and as the travelers advanced slowly broadened into a great open sea churning on the edge of the floating pack. From the top of a hummock Wrangell saw that it extended beyond the skyline. He was granted little time to dwell on his disappointment: the southward setting swell on this "wide immeasurable ocean" was working under the solid ice where he stood, heaving and rupturing it into a web of shattered ice and water lanes, the latter widening rapidly as the blocks of ice floated free. The retreating travelers soon found themselves afloat on an ice island, which was wind driven and shattered on a more solid field, while men and dogs scrambled headlong over bobbing ice pieces to the securer floe on which they had made shipwreck. Wrangell abandoned most of his supplies and made a dash for the nearest land, which he reached with little food for his men, none for his dogs, and at a distance of 250 miles from his nearest depot. Luckily Matushkin's survey party proved to be near at hand, and Wrangell joined it, relinquishing all hope of northern discovery. He had missed no material achievement by his mishap, for what land there was in the Chukchi Sea, Wrangel Island and its tiny attendant, Herald Island, still lay some degrees to the east.

Wrangell had disproved Burney's theory of a land connection with America to his own satisfaction two years before when he had first rounded Skelagskiy Cape. He now obtained evidence which no one could dispute. On April 11, he and Matushkin came upon a pair of rocky bluffs, which rising sheer from the sea and connected to the land by low isthmuses, answered in all points the description of Cape North.* An elderly native of the region recalled as a child having seen out to sea two great ships, which can only have been the *Resolution* and the *Discovery*.

Wrangell carried his survey forward to Kolyuchin Bay and then went back to the Kolyma. There he wound up the affairs of the expedition and together with Anjou returned to Europe.

---

* As mentioned Cape North has been renamed Schmidta, reasonably enough as the old name was confusing and somewhat inappropriate. It is more to be regretted that Burney's Island to the east, off Kolyuchin Bay, is now also called Kolyuchin. This deprives the Arctic historian of a well-deserved monument, and furthermore, with the nefarious application of the name of Dezhnev to East Cape, leaves the eastern Russian Arctic with no memorial to the greatest seaman ever to visit its shores.

A year or two after Wrangell came back from the eastern sector of the Russian Arctic a voyage was made on its western side, short, experimental and without immediate sequel, but of interest as the prelude to several such journeys to be made, by choice or under compulsion, before the century closed. In 1818 the British Admiralty had sent Commander David Buchan and Lieutenant John Franklin with the *Dorothea* and *Trent* to sail over the North Pole to Alaska. The ships were stopped by the Spitsbergen ice and brought back damaged and leaking. This was the last enterprise of its kind, but some years later it suggested to Captain Edward Parry (who in the meantime, like Franklin, had made his reputation in the Canadian Arctic) the expedient of sailing from Spitsbergen to the Pole in flat-bottomed boats fitted with steel runners—amphibious craft, which at need could be hauled up and dragged over obstructing ice fields. The idea was approved and in the spring of 1827 Parry sailed to Spitsbergen in his famous *Hecla* and after much delay in the ice reached the northeastern end of the West Island and berthed his ship in Hecla cove at the bottom of Sorgfiord. From there he set out on June 21, with two boats and supplies for seventy days. The second boat was commanded by Lieutenant James Ross, the destined holder of a record never equaled for the extent of his travels afloat and on foot in northern and southern ice fields. They were escorted for some distance by the ship's cutter under Francis Crozier, another officer of distinction, who was to give his name to a famous penguin rookery in Antarctica before dying in the worst of all Arctic disasters. On the third day out, a hundred miles from the ship the floating pack grew so dense that the boats were hauled up on the ice, and the northward trek begun.

It was soon perceived that all conditions were unfavorable to any significant success. As was natural on this, the first attempt to make a high north on foot, Parry had chosen the warmest months of the year, and consequently suffered handicaps which later explorers tried to avoid—ice fields covered with ponds of water, sharp needle ice, giving poor traction and most painful to the feet, and recurrent leads over which the boats had to be launched and then with much toil hoisted back onto the ice. He was disappointed in the hope of finding a smooth unbroken surface as he advanced to the north. Rains, far heavier than any experienced in the American Arctic were an added annoyance. The ration provided was also too small for the labor of sledging. After weeks of advance Parry became aware of a phenomenon, then novel, but all too familiar to some who came after him: the ice field over which they were traveling was drifting to the south and canceling out a large proportion of the travelers' hard-won gains. An observation on July 22 showed a northward advance of only four miles instead of the eleven that had been logged. On the twenty-sixth the two crews turned back. They had penetrated the ice field to a depth of ninety miles, setting a new "high north" record of 82° 45′ north latitude, in an endeavor that had been bold, original in conception, and deserving of greater success.

More than half a century elapsed between Wrangell's completion of the map of Siberian coastal waters and their actual navigation on shipboard. In the meantime efforts (in which the Englishman, Captain Joseph Wiggins, was prominent) were made to establish a regular shipping route from the west to the Ob and the Yenisey. The Swedish scientist, Baron A. E. Nordenskiold, did much scientific work in that region. To extend these researches further to the east and at the same time perform a popular and sensational exploit Nordenskiold formulated a plan for taking a ship clean through the Northeast Passage and around the continents of Asia and Europe. In January, 1877, he presented the scheme to Oscar II, king of Sweden and Norway, and received His Majesty's approval and promised support. In addition to the backing of the Swedish government, he obtained financial backing from a wealthy fellow countryman, Dr. O. Dickson and from the Russian merchant, A. Sibiriakov. With the funds they provided the 350-ton steamer, *Vega,* built for the northern whaling trade, was purchased and manned with seamen and engineering personnel from the Swedish Navy, under the command of Lieutenant A. A. L. Palander. Nordenskiold took general command of the expedition with a staff of scientists representing several nationalities.

The generosity of his patrons and, in particular the practical help furnished by the merchant Sibiriakov, freed Nordenskiold from two serious difficulties in the fulfillment of his purpose—coal supply and adequate provision for the safety of his crew. The ship could not carry enough fuel for a voyage of such length that was almost certain to be hindered by ice, and not improbably, interrupted by a winter spent icebound in some Siberian bay. Also it was felt that there was serious risk of becoming icebound at the continent's northernmost point, Cape Chelyuskin, where such a mishap could have terrible consequences, for the Taymyr Peninsula was uninhabited and offered no relief or shelter for a shipwrecked crew. Mr. Sibiriakov therefore arranged that two of his vessels, the *Fraser* and the *Express,* should accompany the *Vega* as far as the Yenisey, top off her coal supply and return with commercial cargoes; while a third, the steamboat *Lena,* would accompany her over the hump of the Taymyr Peninsula and Cape Chelyuskin and on to the mouth of the Lena River. From there on the *Vega* could travel alone with a fair measure of security for her crew; for on her right she would have the station of Nizhne Kolymsk and Chukchi settlements, and on the left waters much frequented by American whaling vessels.

On July 21, 1878, the *Vega* sailed from Tromsø for Yugor Strait where she was joined by her three consorts. They started in company on August 3 and reached Port Dickson at the mouth of the Yenisey on the sixth. There the *Fraser* and *Express* parted company and went upstream for their cargoes; the *Vega* and *Lena* went forward on August 10. The passage across the Kara Sea was untroubled by ice but hindered by fog in those uncharted waters where the existence of islands was known, but not their precise location. The two ships reached the west side of the Taymyr Peninsula without mishap and spent August 14–18 in scientific work at Actinia Bay on Taymyr Island. On the nineteenth they put into King Oscar Bay on the west side of an ice-free Cape Chelyuskin. What had been expected to be the principal barrier to success proved to be no obstacle at all.

On rounding Cape Chelyuskin Nordenskiold quitted the land and set an easterly course to determine if any new land lay between Cape Chelyuskin and the New Siberian Islands. But the ice now began to grow dense and in the fog still prevailing a way could hardly be found through the floating masses. Mindful of the fate of the *Tegethoff* and not disposed to risk the future of the voyage for chance discoveries Nordenskiold drew back and followed the land down the shore of the peninsula and then eastwards. Southeast of Semenovskiy Islet (where three years later a disastrous gale was to overtake the boats of the American *Jeannette*) he delivered his mails to the attendant *Lena*, which turned shorewards to go up the Lena to Yakutsk. Prevented by ice from landing on Lyakhov Island for scientific studies, Nordenskiold pushed on with the confident hope of breaking out into the Pacific before the freeze. Svyatoy Nos was passed on August 31, the mouth of the Kolyma and the Bear Islands on the third. From there on ice began to move in, restricting the *Vega* to a channel with the pack on her left and to the right land where newly fallen snow lay unmelted on northern slopes. On the fifth Baranov Rock and Chaunskaya Bay were left behind, but now young ice was becoming an obstruction. To the east of Cook's Cape North, reached on the 12th, ice pushed up into shoal water compelled the *Vega* to anchor in the lee of a grounded floe until the eighteenth when south winds drove the pack off shore and reopened a navigable channel. For ten more days she labored on with numerous checks and hindrances. On the twenty-seventh she made a circuit deep into Kolyuchin Bay to round the wedge of pack in its mouth. The next day ice driven right against the land brought the season's voyage to an end.

The *Vega* was frozen in at 67° 4′ 49″ north latitude, 173° 23′ 2″ west longitude, a mile from the land. Cape Serdze Kamen lay forty-five miles to the east and beyond it, 120 miles from the ship, was East Cape and an unobstructed passage home. It was mortifying for Nordenskiold to be held up at the very last moment on the brink of a stunning success, but only good fortune had brought him so far. Judged by other polar voyages of the period, or even by the performance of high-powered steel-encased ice-breakers of a later period, this bold venture had been attended by luck.

As the ship lay in the open sea behind no other breakwater than floebergs piled up in shoal water, a large depot was placed on land as insurance against shipwreck. It was not needed. No worse evil befell the men of the *Vega* than the tedium insepa-rable from nine months of captivity on a polar shore. Some relief was found in visiting back and forth with the Chukchi of the neighborhood. Their chief, who held some sort of appointment under the Russian government, even undertook the delivery of the ship's mail to the nearest post office. Thus the *Vega*'s situa-tion became known to the world, a circumstance which was to have some influence on the fortunes of the *Jeannette* in the en-suing August. Nordenskiold himself did not look for release until August, but the unpredictability of the ice which had caused his captivity brought him an earlier departure than he had looked for. On July 18, 1879, the sea ice parted from the land opening a channel of which the men of the *Vega* were not slow to avail themselves. Cape Serdze Kamen, long a close neighbor, was left behind on the nineteenth; on the twentieth the ship passed southward through Bering Strait to begin "the remark-able triumphal procession from Japan to Stockholm, which stands unique in the history of festivities." Yokohama, Hong Kong, Cairo, Naples, London and Paris were some of the places where Nordenskiold was feted before arriving on April 24, 1880, for a series of grand receptions at Stockholm.

THE AUSTRIANS: VOYAGE TO FRANZ JOSEF LAND

For half a century after the journey of Wrangell, scientific enterprise in the Russian Arctic concerned itself more with the exploitation of known regions (especially in Novaya Zemlya) than with the discovery of new land and the development of new areas of research. The last third of the nineteenth century, however, saw an expansion of knowledge into areas before unknown. The stimulus for this came from a country little disposed by tradition or geographical position for polar discovery. Two young Austrians, Julius Payer, a graduate of the Austrian military academy and an experienced mountain climber, and Karl Weyprecht, a naval lieutenant, emerged as promoters of a national polar expedition for oceanographic and other scientific purposes, and with the generous patronage of Count Wilczek and others were enabled to put their purpose into effect.

The Gulf Stream makes the eastern Russian Arctic more accessible to shipping than corresponding latitudes farther west. Thus, owing to its comparative approachability from a Norwegian port, Payer and Weyprecht were able to carry out what would have been less practicable in the American Arctic waters, a preliminary voyage of reconnaissance. Intending to study the influence of the Gulf Stream, temperatures in ice and water, ice conditions and magnetism, they chartered the fifty-ton Norwegian *Isbjorn*, and in June, 1871, sailed from Tromsø to Bear Island. There the ship came under heavy pressure and was barely saved by a protecting cushion of brash ice. After calling at

Cape Lookout on South Spitsbergen the explorers steered east to Novaya Zemlya. On this course they twice pushed up beyond latitude 78°, but were disappointed in their quest for a north-ward trend in the ice edge. They coasted homewards down the east shore of Novaya Zemlya, confirmed in their intention of making it their point of departure in their northward thrust.

The main expedition in the *Tegethoff* sailed from Bremer-haven, Germany, on 13 June 1872. Payer was the overall com-mander; responsibility for the management of the ship rested with Weyprecht. Other officers were Lieutenant Brosch and Midshipman Orel of the Austrian navy, Dr. Kepes, surgeon, Captain Olaf Carlsen, ice master and harpooner. By what was to prove a lucky inspiration, Payer had enrolled two old com-rades, the Tyrolese mountaineers, Haller and Klotz. The crew was a confusion of nationalities—German, Slavonian, Italian and Norwegian: orders were given in Italian. One officer was for a time constrained to converse with Italians in German and with the Norse Carlsen in English, but undoubtedly communi-cation grew easier in the course of the next two years. Total ship's company numbered twenty-four. They were furnished with a *ukase* (decree) in which the Russian tsar required his subjects to furnish the travelers with any assistance they might require.

Leaving Tromsø on July 13, the *Tegethoff* set a northeast course for upper Novaya Zemlya. The season proved to be ab-normally bad; the ship ran into heavy ice only seventy miles north of Matochkin Shar and could only advance by the unsafe and disagreeable expedient of hugging the coast on her north-ward course. Count Wilczek was met in the *Isbjorn*, come to plant an emergency depot on the Barents Islands near Cape Nassau. This was effected on August 15 and for a week after both ships lay there, icebound and motionless. On the twenty-first the ice pressure relaxed, and the *Isbjorn*, with her mission accomplished, lost no time in extricating herself and steering a course for home. The *Tegethoff* set off in the opposite direction, but in a few hours was again imprisoned, in 76° 22′ north latitude, 63° 3′ east longitude—this time permanently. As she drifted northwards in the grip of the ice the idle crew sought occupation in hunting polar bear and building ice houses. On October 2, they crossed 77° north latitude and the shores of Novaya Zemlya slowly faded from sight. On the thirteenth, the ship suffered her first nip to the tune of "shrieks and howls" from the grinding ice. These ice pressures recurred daily in varying degree for more than four months. In spite of the supply of bear meat, traces of scurvy occurred. Bathing was found wholesome, but added too much to the dampness on board. Schools, four of them owing to the babel of languages, were operated, and morale well maintained. The seamen boasted that they were wintering farther north than any ship in history, unaware, of course, that the season before the American, C. F. Hall, had berthed his *Polaris* at an even higher latitude in Thank God Harbour on the West Greenland shore. The ship, until then immovable, began to drift east in January: at the end of February she was in 79° 12′ north latitude and about 72° east longitude. She then began the devious but generally westerly drift which was to carry her to new land.

At the end of April the ice was seen to be rifting, raising hopes of an early liberation. Attempts to saw and blast the ship loose were without success, for successive pressures had lifted her up and jammed twenty-seven feet of ice beneath her keel. As the surface melted away the ship remained perched in the air and was only kept upright by means of shores which were daily adjusted as the ice on which they rested rotted and gave way beneath them. In mid-July the *Tegethoff* began to eddy around south, east and then north, a movement explained by the nearness of the land which they were soon to discover. In August she was still held captive and seal hunts were organized as a precaution against scurvy during the second Arctic winter which was growing daily more probable.

At midday, August 20, 1873, 79° 43′ north latitude, 59° 33′ east longitude, a rising mist revealed "outlines of bold rocks" and slowly unveiled a lofty coast with a bold alpine profile and crowned with glaciers, the source, obviously, of the numerous icebergs recently seen. Delighted as they were at this find—which in a moment changed their seemingly futile cruise into a sensational success—the Austrians dared not attempt to make land over rotten and rifted ice: they had the mortification of seeing the newly discovered land receding as the moving ice carried them away to the south. Had this drift continued they would have been hard put to convince the world that their find was genuine: luckily they eddied back to 79′ 58°, the highest latitude the ship was to attain, to the east of Wilczek and Salm Islands and just south of Hochstetter Islet. A bold attempt by Payer and five others to reach the latter (for the ice was still in motion and they might never recover the ship) was thwarted in the foggy autumn twilight. But for the guidance of a dog they might have been lost forever. The still grinding ice wore down the floe in which the ship was imbedded and carried her close to a headland on the south side of Wilczek Island (not to be confused with the Wilczek *Land* which Payer later placed on his map). There the *Tegethoff* remained firmly locked in by the winter frost. In November brief visits were made to the south shore of Wilczek Island, but systematic discovery was deferred until the return of daylight.

Beyond Wilczek Island the Austrians discovered two larger lands which those generous cosmopolitans named after the contemporary British and American discoverers, McClintock and Hall. The island group received the name Franz Josef, after the ruling Austro-Hungarian emperor. The two commanders resolved to add what they could to discoveries already made by an extended spring journey and then to abandon the ship and make for Novaya Zemlya with sledge and boat.

On 10 March 1874, Payer set out with a sledge party on a journey to McClintock Island. Daylight had returned and "the level surface of the young ice glowed with the colours of the morning." Less agreeable was a rising wind sweeping over the ice fields, and the sight of a polar bear dogging their footsteps through the driving snow. For security they turned back and killed the brute, for there was a danger that he might come close upon them in a blinding snow squall before a shot could be fired. Finding himself cut off from McClintock Island by shattered ice Payer turned back to Hall Island, passed under the towering Cape Tegethoff and climbed the cliff on the island's southern face to a height of 2,000 feet. With his trusty *jäger* (hunter), Haller, he went on up to the top of Sonnklar Glacier and to the northwards made out what he took to be large land masses, Zichy Land to the left and Wilczek Land to the right, divided by the broad belt of Austria Sound. These were to be broken up into clusters of islands by later discoverers.

The trip to McClintock Island was preliminary to the main effort, a sledge journey of a month's duration directed northwards up Austria Sound. Plenty of courage was required to undertake it, for owing to the pressure of the moving pack against the islands, disruptions in the ice could occur early in the spring season; therefore, there was no guarantee that the returning travelers would be able to reach the ship, supposing that she still lay at Wilczek Island to receive them. Payer set out on March 26 with the midshipman Orel, the *jägers*, Haller and Klotz, three seamen, along with a team of dogs.

Rounding Cape Frankfort on the east of Hall Island and noting the Willersdorf Mountains on Wilczek Land to the east, the party made rapid progress northwards passing Cape Tyrol on Wiener Neustadt Island in a blizzard without then being aware of it. As the storm subsided the bulk of Crown Prince Rudolf Island began to loom ahead. At first Payer believed that Wilczek Land to the east terminated about this latitude; but the mists began to rise and the glare of the sun on rounded fog banks gave him the illusion of "glittering ranges of enormous glaciers" (Dove Glacier) reaching north to Cape Buda Pest.

These semblances were visionary and their presence on the chart was to cause much perplexity to Nansen when, under a clearer August sky, he approached the archipelago from the northeast.

Finding Rawlinson Sound (dividing Crown Prince Rudolf Island from the supposed Cape Buda Pest) a labyrinth of hummock and loose snow, Payer swung west under the cliffs of the island and camped on Hohenlohe islet. From there he determined that the ice cap of Crown Prince Rudolf Island was 3,000 feet high and to be mounted with light dog sledge only. He therefore left Haller and two seamen at the camp and went on with the rest, lightly equipped to climb the steep glacier face of the island's southern side. In making this ascent the seaman Zaninovich with sledge and four dogs broke through the snow bridge of a crevasse and plunged down to an ice shelf thirty feet below. Payer was caught and dragged to the ground; he freed himself by cutting a trace. It was quickly ascertained that Zaninovich and dogs were alive and not seriously hurt, but their comrades had neither the strength nor the means to rescue them so Payer tramped back six miles to Hohenlohe Island and brought up Haller and the others with ropes and tent poles. Haller, the mountain climber, was let down into the crevasse. He found the sailor barely alive by that time, but conscious; the dogs, with the philosophy of their breed, had slept through the period of crisis. Man and dogs were hoisted up, after which Haller unpacked the sledge and delivered it and its freight piecemeal to his comrades above. Payer sent back the extra men, descended again to the sea and began to round the island by west and north, noting Karl Alexander Land to the southwest. He rounded Cape Auk, and passed the later notorious Teplitz Bay with icebergs to seaward, looming gigantic through the mist. As the ice was newly formed and unsafe, the men advanced roped together as if mountain climbing. Water oozing through the fractured ice slowed the advance of the weary travelers. Approaching Cape Saulen they found that tide and current had shattered the ice to the very foot of the cliffs, and so ended their journey by sea. A glacier offered a convenient ramp; they went up this and passed along the cliff top at an altitude varying from 1,000 to 3,000 feet. On April 12 finding themselves on a glacier crisscrossed with crevasses, they left behind sledge and dogs and walked on a few miles to Cape Fligely. They reached the point where the land of the Eurasian sector approaches most nearly to the Pole. Payer did not suspect what he had accomplished. He thought that he could see the land sloping away northeast to Cape Sherard Osborn and, beyond that illusive cape, a separate land which he put down as Petermann Land. The advance party was now a hundred and thirty miles from the ship. They quickly covered the forty miles back to Hohenlohe Island where their weaker comrades were still encamped.

On the way out the explorers had been spurred by hope and the sense of daily achievement. With this stimulus gone they began to be vexed with a gnawing anxiety about the ship, which frequent evidence of disturbance and rupture in the sound did nothing to allay. The men were tired and low in morale. Payer tried to encourage them by recounting the daily mileages which McClintock had registered in the Canadian Arctic, with the success with which such admonition is usually repaid: some listened with indifference, others with open scepticism. As they trudged down Austria Sound, the leader and his trusty Haller continued to quit the line of march and scale mountains for further observations. The main party was much hindered by "deep layers of snow, and great rents in the ice caused by the falling of icebergs." Haller, the pioneer, plunged through and barely escaped drowning. Thereafter they advanced with increased caution, testing the surface with a sharpened pole. The men were sick of "the animal labour of dragging" and bitterly commented on the appearance of a "water sky" to the south and the mutter of distant surf. A little to the south of Wiener Neustadt Island they arrived at the shore of a vast expanse of open water which cut them off from Hall Island. A cache left on the ice earlier had been carried away; they were virtually without provisions and fifty-five miles from the ship.

In this extremity Payer turned to the east and after a tramp of fifteen miles found to his relief that the ice edge continued straight to the east and ended in the cliffs of Wilczek Land. They mounted these and groped southward over a heavily crevassed glacier where the howling of a blizzard mingled with the roar of the surf many feet below. The carcase of a bear, killed on the outward journey, was found and removed the danger of collapse through hunger. The vast lake which they were skirting was found to terminate at the mouth of Austria Sound. They crossed to Cape Frankfort, passed between Salm and Wilczek Island on firm ice and with unspeakable relief saw the slender masts of the *Tegethoff* rising above the hummocks which surrounded her. All were safely aboard on April 22, completing a journey less remarkable for time and distance than for the novelty of its setting and for the uncertainties and hazards which it entailed.

A week later the unwearied Payer paid a visit to McClintock Island, climbed to its summit and studied the medley of islands to the north of Markham Sound. To him they still appeared to be fused on the solid mass of Zichy Land. "It often happens," he observes elsewhere, "that banks of fog on the horizon assume the character of distant ranges, for the small height to which these banks rise in cold air causes them to be sharply defined." When he visited Franz Josef Land Nansen confirmed this observation, and, though he helped to expose the Austrian's errors, thought them amply excused.

It had long been determined to quit the ship when the spring explorations were finished. In the tideswept waters where she lay the *Tegethoff* might well break loose in midsummer, but the risk was not worth taking. The fate of Franklin was an ever-present reminder of what might happen to those who trusted to such a chance. It was proposed to take sledge and boat to the depot planted by Count Wilczek on the Barents Island, 200 miles away and thence to coast down the Novaya Zemlya shore, trusting to rescue by Russian fishing boats. They deserted the ship on May 20, 1874.

The distance to be traveled, though formidable enough, was in itself the least of the obstacles to a happy end to the journey. The strength of the entire crew was required to drag each of the three boats. Advance was made in three relays, and this by a devious course around hummocks, through moist and spongy snow concealing jagged ice boulders, still frost chilled and as hard as steel. As they drew away from the land they found that the ice was circling in a vast eddy which more than once bore them back to within sight of the abandoned *Tegethoff*. After five weeks the net southward gain was forty miles only. Then on July 4 a south wind drove them up to latitude 79° 40'. After two months of "indescribable effort" they were only two German miles from the ship. A return to it for a third winter was seriously debated. On July 16 Cape Tegethoff on Hall Island was still clearly visible. In the latter part of the month real progress was made on foot, augmented by the occasional opportunity, always gratefully embraced, of traveling afloat in a southward trending lead. On August 7 the ice underfoot was felt gently rocking to the ocean swell. After a halt of some days for caulking and general repair on the boats, the little party resumed its march and on the fifteenth emerged from the jungle of hummock in which it had strayed for three months and saw to the south the level surface of an open sea.

They embarked at about latitude 77° 40'. With oar and sail progress appeared to be miraculous. At noon on the seventeenth "silvery points," the mountains beyond Cape Nassau, rose above the level of the sea. Then fog came down so low and dense that the boats seemed to be floating in clouds. Though the current was sweeping them past the depot at which they had been aiming they had not the patience to halt or turn back. Fortune, however, was still disposed to be grudging in her favors. When the fog lifted the crews turned in shore to land and ease stiff and swollen limbs. Two years before level shore ice had offered easy access to the land. Now no ice was to be seen but a heavy surf breaking on bare, unapproachable cliffs. On the twenty-first they arrived at Matochkin Shar and were sadly disappointed to find in this well-known and convenient harbor none of the fishing vessels on which they had reckoned. Fifty miles further on lay the much frequented Bay of Dunes. Failing a ship there nothing remained but the desperate expedient of crossing the Barents Sea to Lapland, 450 miles distant. In view of this possibility and of the consequent likelihood of parting company in a storm Payer made an exact distribution of what rations were left among the three boats of his command.

On the evening of the twenty-fourth they were still coasting down the Novaya Zemlya shore when a shout was raised from the leading boat. A skiff was sighted ahead with two men apparently hunting seafowl. They soon opened up a bay where two ships lay at anchor and soon made fast alongside the Russian *Nikolai.*

Her patriarchal captain, Feodor Voronin, received them with a heartfelt kindness, no whit lessened by the formality of his manner. He bared his head and bowed at the sight of the Imperial *ukase* by which he was enjoined to give the bearers all the help in his power. The spontaneous hospitality of the men equalled that of their captain. One poor seaman observing Payer silent and seemingly depressed, made him a gift of white bread and tobacco. "Though I did not understand a word he said his address was full of unmistakable kindness, and so far needed no interpreter." Dr. Kepes, invited to give treatment to a man on the other ship, came back laden with tobacco for himself and his comrades.

Captain Voronin's intention of remaining another fourteen

days on his station was most unacceptable to the Austrian crew which had had its fill of delay and frustration. With a gift of boats and rifles and promise of further reward they induced him to sail immediately and so ended their cruise in safety at the Norwegian port of Vardo.

THE *JEANNETTE*: ADRIFT IN THE ICE FIELDS

Payer's foray into the upper latitudes of the Russian Arctic, though not immediately followed by the adventurers of western Europe, was the beginning of a new epoch in polar discovery. Within five years a second deep penetration was made, this time from the eastern flank by way of Bering Strait.

The plan for this cruise originated with Lieutenant George Washington De Long, U.S.N. who as an officer of the *Juniata* had played a distinguished part in the search for the lost crew of the Arctic ship *Polaris*. He obtained the backing of the United States Navy and of James Gordon Bennett, proprietor of the New York *Herald*, who had financed the first African journey of Henry Morton Stanley. The navy undertook to direct the expedition and to provide personnel, all under naval discipline; Mr. Bennett assumed the burdensome and less spectacular office of paymaster.

The objective of the expedition was the North Pole. To achieve it De Long was constrained to test an area of approach hitherto untried—the only one left him. Sir Edward Parry had failed by a large margin to reach the Pole from Spitsbergen; Payer had done no better in the western Russian Arctic; the German Koldewey expedition had come to grief on the east coast of Greenland. And while De Long's plans were taking shape, the British expedition of Sir George Nares demonstrated the inaccessibility of the Pole from the eastern American zone—at least by the methods of travel then employed. There remained only entry by way of Alaska, and De Long cherished the hope that the warm Japanese current, passing through Bering Strait and welling out into the Arctic, would melt the ice and open the sea for navigation to a higher latitude than could be attained elsewhere. This was a glaring rationalization—apart from being sprinkled with islets the strait was narrow and too shallow to give passage to a current of any force or volume. It may well be that De Long used the warm-water theory to reassure his backers, while placing his own reliance on other more practical considerations.

In 1849 Captain Henry Kellett of H. M. S. *Herald,* on the lookout for Franklin's missing ships, had entered the Chukchi Sea, neglected since Wrangell's time, and had made a landing on the tiny Herald Island, 200 miles north of Cook's North Cape.* He was hindered by ice from approaching another land dimly visible to the west. In 1855 Captain John Rodgers of the U. S. S. *Vincennes* called at Herald Island and also made out land to the west, but so dimly that geographers were sceptical and unconvinced. Their doubts were soon resolved. American whalers became active in the Chukchi Sea, and in 1867 Captain Thomas Long of the ship *Nile* obtained a clear view of the new land, and from a distance of fifteen to eighteen miles traced the entire length of its southern shore. He was unable to form any notion of how far his discovery, named by him Wrangel Land, extended to the north. In this uncertainty the geographer, Dr. Petermann, revived the theory of a Bolshaya Zemlya, and formulated the theory of a saddle of land reaching over the top of the globe and linked with Greenland on the other side of the Pole. Such a hypothesis was not improbable and most attractive to De Long: a northward-extending shoreline would afford the explorer better prospect of navigable waters and, failing that, chances of a secure harbor for his ship; it would provide

* Sir Henry Kellett, K.C.B., could not have foreseen that this apparently innocent transaction would earn him denunciation as an "*American* spy" by a Soviet historian of the Stalinist epoch. L. M. Starokadomski, *Pyot plavaniy b severnom leditom okeane, 1910–1915.* (Moscow, 1959).

advancing sledge parties with safe bases for their provision depots, and a surer road than the moving ice which had so distressed the parties of Parry and Payer. So De Long's project was by no means visionary: he could, without extravagant optimism, hope to emulate his fellow countryman, Dr. Kane, and open a new and promising line of approach to the mysterious regions above 80° north latitude. Instead it was to be his achievement to reduce the "Great Land" of Siberian myth to the present modest proportions of Wrangel Island.

As Bennett's agent De Long purchased in England the barque-rigged steam yacht *Pandora* (later rechristened the *Jeannette*) and brought her to the San Francisco dockyards to be made ready for Arctic service. He spent much of the period of the refitting at the Navy Department in Washington, leaving to his first officer, Lieutenant Charles W. Chipp, the task of selecting a crew. Like C. F. Hall, Chipp found it difficult to enlist native Americans, and in this connection received the following directive from his commanding officer: "Norwegians, Swedes and Danes preferred," wrote De Long; "avoid English, Scotch and Irish. Refuse point-blank French, Italians and Spaniards." The motive for this somewhat arrogant distinction was to secure men whose heredity and early environment seemed best to fit them to withstand the rigours of an Arctic winter. De Long made himself responsible for the appointment of officers, and it proved fortunate that his ban on Scots did not apply in this category. As chief engineer he chose Passed Assistant Engineer George W. Melville, grandson of "James Melville of Stirling," the destined hero of the expedition. Lieutenant Chipp had already assumed the duties of executive officer and second-in-command. Master (later Lieutenant) J. W. Danenhower was appointed as navigating officer and Passed Assistant Surgeon J. M. Ambler as surgeon. In addition two civilians, Messrs. Jerome J. Collins and P. L. Newcomb, were enrolled respectively as meteorologist and naturalist.

The *Jeannette* sailed from San Francisco 8 July 1879, but she was so deeply laden that she made slow going to the north and was held up off Alaska by the delayed appearance of the supply ship that was to top off her coal. She passed Bering Strait on August 28, but lost more time by making inquiries—in conformity with her orders—for the still unreported *Vega* of Nordenskiold. Informed by the natives at Cape Serdze Kamen that the *Vega* had broken out and sailed for the Strait several weeks before, De Long headed north from the Siberian shore and set a course for Wrangel Land. On September 4 Herald Island was sighted. And then, as a climax to two months of delay and frustration, the ship was firmly and permanently beset. On the fifth she began to travel through the pack; on the sixth she was icebound; on the seventh she was nipped and heeled over five degrees. Never again in the twenty-one months that elapsed before her destruction was she able to steam her own length. (Stefansson's *Karluk* was to come to grief near that spot thirty-five years later.)

In the winter of 1879–1880 the *Jeannette* was pushed back and forth by wind and current, tracing an intricate pattern in the ice fields to the north of Herald and Wrangel Land. The movement was sufficient for Lieutenant Chipp to take a number of observations and roughly to measure the length of Wrangel Land, which it was the *Jeannette*'s achievement to reduce from a supposed continent to an island of moderate proportions. On January 19, 1880, a bad nip started a leak in the ship's bow. This was brought under control by the exertions of Melville, the carpenter Sweetman, and the quartermaster Nindemann, but from then on continual pumping was required to keep the ship afloat. The anxiety of the captain was aggravated by the sickness of Lieutenant Danenhower, disabled by an affection of the left eye and threatened with loss of vision in the right. A surgical operation was required to avert total blindness. Thereafter Danenhower was never able to resume the routine duties of his rank, though he was to render important services in more crises than one.

The Jeannette: *Adrift in the Ice Fields*

The success denied the crew of the *Jeannette* was more than merited by the conduct of officers and men. Helpless in the grip of the ice for more than one and a half years, and in a state of inaction—of all things most damaging to morale—they preserved health, good humor and discipline. The leak, controlled, but never perfectly repaired, may not have been an unqualified evil, for it occupied attention and gave employment. Polar bear, seal and walrus were hunted with success. An outbreak of scurvy was averted by the constant attention of captain and surgeon. As Dr. Ambler was never satisfied that the snow or ice was low enough in salinity to warrant use for drinking, water was distilled over a coal furnace. As this fuel was also used to operate the pump and keep the ship afloat, it was in uncomfortably short supply as the sun receded and the probability of a second winter in the ice pack hardened into certainty.

Wrangel Island had been lost to sight on March 24, the ship having been carried some eighty miles to the northwest since being beset to the east of Herald Island. She passed the summer weaving back and forth in a restricted space. In mid-August she was perhaps sixty miles to the north but actually to the east of the position of March 24. Only toward the end of 1880 was she caught in the steady westerly drift, knowledge of which was her second contribution to the lore of the Arctic. Observations on January 4, 1881, showed a drift of 38½ miles north, 58 west, since December 31, a rate slackened, but hardly ever reversed until the ship had traveled 400 miles in a west-northwesterly direction. De Long did not live fully to appreciate the impulse to discovery he was giving, but he was cheered by his progress and hoped soon to reach navigable waters where the outflow from the Lena had dispersed the ice, or else—he was haunted by the ancient vision of the "Open Polar Sea"—of finding open water in the high north to which the direction of the drift was slowly conducting him. Though not despondent, he was naturally impatient: "Nothing but ice, day after day," he wrote in his journal. "Hummocks, large and small, ridges high and low, a rough tumbled mass over which there is no path, and through which there is no road, and in the centre of the picture a poor little ship buried to her rails in snow-drifts,—a stranger in a strange land, indeed! As day adds to day the sameness becomes wearying, and after long experience of it, maddening."

This entry was made on April 25, 1881. Three weeks later events began to happen to move rapidly toward catastrophe. On the evening of May 16 ice pilot Dunbar, an old whaler, reported land in sight to the westward: "Our voyage, thank God, is not a perfect blank." In the ensuing days the drift carried the ship past the northern face of the new land at a distance of thirty miles. It was a small rocky outcropping with snow-filled clefts and, no doubt, the resort of bears and wildfowl. It was named Jeannette Island. On the twenty-fourth a similar outcropping was reported fifteen miles to the west of the first discovery. This was named Henrietta after a sister of Mr. Bennett. The next year A. W. Greely was to give the same name (in honor of his wife) to a glacier in the Lake Hazen region of Ellesmere Island.

On May 30, as the drift had brought the ship to within an esti-mated distance of twelve miles from Henrietta Island, De Long resolved to disregard the dangers of rotting ice and opening leads and send a party to land. The supremely competent execu-tive officer, Lieutenant Chipp, was temporarily, and Mr. Danen-hower permanently, on the sick-list, so the mission was entrusted to Chief Engineer Melville, along with Mr. Dunbar, Ninde-mann, an old *Polaris* hand, and three others. They came back to the ship on June 5. Melville reported that they had reached the island on June 2 and examined it as much as the twenty-four hours allotted for their stay permitted. It proved to be "a desolate rock, surmounted by a snow-cap, which feeds several discharging glaciers on its east face." Steep cliffs had confined the party to the beach. They found no game except dovekies, and no other forms of life save moss and grass. The journey had been made terribly severe by the frequent necessity of road building and ferrying over leads, with the consequent necessity of unloading and reloading the sledge cargo. The dogs, panic-stricken on insecure ice, had given much trouble to their inexperienced drivers. Some may have guessed that all this was only a prelude to a far longer journey—for most of the crew their last.

Had Melville's journey been deferred for another ten days he would have had no ship to come back to. The officers had already debated the advisability of deserting the ship and making for the Lena Delta, seven hundred miles to the southwest, using the New Siberian Islands as a stepping-stone on their way to the continental shore. Their commander was now relieved from making this difficult decision. For some days after Melville's return De Long noted a disturbance and uproar in the pack that was natural enough when the moving ice field was pressing against land. At midday of June 10 the ice around the ship opened up and she slid into the water on an even keel. The rudder, unshipped since the autumn of 1879, was rehung in anticipation of complete liberation; in the meantime the ship was moored to the ice on her starboard side. On the morning of the twelfth the ice barrier to port was observed to close in a little. Movement began again with great force at 4 P.M.: the ship was brought under pressure and heeled over 22 degrees, her port bow thrown up; while stern and starboard quarter were tightly gripped, the spar deck buckled, and the starboard side burst open. At 6 P.M. the ship began to fill. Boats, food and equipment were got out onto the ice and at 8 P.M. the order was given to abandon ship. At 4 A.M. on the thirteenth the *Jeannette* righted to an even keel and sank from sight.

The predicament of the shipwrecked crew is best appreciated from the standpoint of the civilized world which they had left so far behind. Of all regions where life is found, the Siberian coast to the east of the Lena is the coldest, the most barren and the least accessible. A hundred miles north of this forbidding shore lay the large group of the three New Siberian Islands, bare rock, neither inhabited nor habitable, and unvisited for the last sixty years. De Long and his men had been cast away at a somewhat greater distance to the northeast of these islands (at 77° 15′ north latitude, 154° 59′ east longitude), in the heart of the ice. It was the worst season for travel in the Arctic for the surface was neither hardened by frost, as in the spring, nor drained of the summer melt, as in autumn. The way to safety lay through a maze of puddles, over rotten ice and moisture-laden snow, all which provided bad footing and worse traction for sledge runners. Rations had to be carried for thirty-three men on a journey of uncertain duration, but sure to be slow, and worse, boats also had to be dragged on heavy sledges until open water was arrived at. As De Long well knew this task had broken down Markham's crews in the Canadian Arctic five years before. He also foresaw that the ice, while it lasted, would not offer an uninterrupted passage, but would be broken by leads impassable except by bridging or ferrying. A further handicap was the incapacity through sickness of his "line" officers, Chipp and Danenhower. Hence to supervise the complex task of surveying a route, hacking a road through tumbled ice masses, and shuttling back and forth to bring on boats, supplies and the sick, he had to rely on Mr. Dunbar, a fine old sea dog but too advanced in years for footslogging, on Chief Engineer Melville, and Dr. Ambler. The record of naval surgeons in the polar travels of the nineteenth century is a glorious one. Two of the best known, Sir John Richardson and Sir Alexander Armstrong, who ended their days full of years and honors, hardly exceed in merit the Virginia born surgeon who, while yet in his prime, died of want in the Lena Delta.

Unflurried by his critical situation De Long spent several days in sorting out provisions (of which he was taking a ninety days' supply), assigning loads, and making ready the sledges. Three boats were to be taken, the first and second cutters and the whaleboat. The order of march was as follows:

> 1st. All hands drag the first cutter. Dogs drag the No. 1 sled.
> 2nd. Starboard watch drag the second cutter. Port watch drag the No. 4 sled. Dogs drag the No. 2 sled.

3rd. Port watch drag the whaleboat, Starboard watch drag the No. 3 sled. Dogs drag the No. 5 sled.

That is, five miles of travel, three of them burdened, for one mile of distance gained.

Being well aware that the heat of the sun would be barely tolerable to laboring men, though the temperature of the ice surface was near freezing, De Long resolved to rest by day and journey in what were technically the hours of night. They started at 5 P.M., June 18, on a march which not surprisingly was attended with confusion and, one suspects, with some pardonable ill temper. Mr. Dunbar had planted four flags to mark the road ahead; De Long could see only three. He ordered the provision sledge to halt at the third flag, while Melville went on to the fourth, a stage beyond the intended encampment. To add to the confusion the sick were advancing slowly; they and some of the sledges were cut off by an opening rift in the ice with consequent ferrying by dinghy and much dragging and heaving of sledges and cargoes. Progress for the first day was one and a half miles, and De Long noted sourly that for some days any deficiency in supplies could be repaired from the dump of surplus stores lying still visible on the ice where the ship had gone down.

One or two such days would make the *plan* of march clear to all who shared in it; its *execution* was hindered by what no amount of understanding could control. Runners of overburdened sledges would fold under, necessitating a halt, unloading and repair; men would lose footing and plunge up to the neck in icy water. A hot sun alternated with damp, blinding, depressing fog and cold and rain—the least bearable of these afflictions: "When a rainy day sets in, one's misery is complete. Even the dogs cower under the boats like hens, or snuggle up against the tent doors begging for admission." It was observed however, that rain was usually followed by a sharp frost and a greatly improved surface.

At midnight on June 25, after a week of travel, De Long obtained a meridian sight and found his latitude 77′ 46°—which meant that moving ice was carrying his men three miles north for every mile south made on foot. He hoped that this result, obtained when the sun was low, might be due to refraction, but a noon observation of 77° 42′ more or less confirmed it, placing the party twenty-eight miles north of their point of departure. The commander composedly noted, "Instead of making a south course I shall incline more to the southwest, for as the line of our drift is northwest, a southwest course will cross it more rapidly than a south one and bring us quicker to the ice edge." De Long disclosed the fact to Melville and Ambler, whose rank and zealous exertions gave them a claim on his fullest confidence, but dared admit no others to the secret for fear of its reaching the men: "I dodge Chipp, Danenhower, and Dunbar, lest they should ask me questions." This contrary drift was the fatal mischance which was to cost most of the party their lives. It might have made all the difference had they reached the Delta two weeks earlier with rations in hand.

On June 26, five leads in succession required bridging with a supposed gain in distance of one and a quarter miles in eleven hours. The twenty-seventh was even more trying:

> Just after leaving our halting place, we had another opening to cross twenty feet in width; and while we tried bridging it, it opened twenty feet more. After great exertion we succeeded in dragging in three large floes for bridges, and by herculean efforts got our sleds and boats over, launching first and second cutters. Drifting about one eighth of a mile further, we had another ice opening about sixty feet wide, and to bridge this we had literally to drag an ice-island thirty feet wide and hold it in place. Hardly had we done this when the lead widened, and we had to scour around for more huge blocks to make them serve our purpose.

Though fearful obstacles to foot-passengers, these rifts gave no immediate prospect of travel by boat. Later De Long was to observe that the leads always ran east-west, and whatever their length were of no help to a boat party whose course lay south:

"It is hardly late enough to find leads of any length, but there are openings enough to give us serious trouble." To make matters worse the men complained of the heat with the thermometer at 30° F: "It seems curious enough to see men seeking a shady spot in which to sit and smoke while the temperature is so low." On the twenty-eighth the captain had leisure to admire the beauty of cloud formations. He was positively cheered by the more frequent occurrence of the loathsome fog, as a sign of near approach to open sea. July 3 found them at 77° 31′ N, 150 E still fifteen miles north of the latitude in which the ship had gone down.

During the next week twenty-six miles of positive gain was registered in a direction (owing to the vagaries of the ice) a little east of south. On July 10 the men asserted that land was visible to the west, a belief which the captain, sceptical at first, confirmed by his observation on the eleventh. It could not, he felt sure, be New Siberia, but it lay more or less in his course and offered a relatively dry and secure resting place. It proved harder to reach than he foresaw.

About this time Mr. Chipp returned to duty and replaced, as chief transport officer, Melville, who in turn took over from Dr. Ambler charge of road building and bridging. The doctor, says De Long, became a reserve—in effect the assistant of Dunbar as pioneer and surveyor. The ice pilot no doubt was glad of help, for his problems were multiplied by the erratic movement of crumbling ice pieces, while the boats were too battered and leaky to be a safe conveyance until repaired. They were used for short traverses, and the labor of moving cargoes was sometimes avoided by working the loaded sledges on board the cutters for transfer. Vaulting over a lead De Long plunged into the water and was hauled out by Dunbar, who "grabbed me by the hood, as he thought, but by the whiskers principally, as I realised, for he nearly took my head off." July 21 was a day of rain squalls with a moderate northeast gale. "The confusion was such that I dared not try to cross anything. Large blocks, small lumps, and floebergs were moving along the southward, and occasionally a large piece, seemingly free, would suddenly be shot up in the air as it was squeezed by a larger one, or its submerged portions became freed from overriding masses." In calmer weather they went on, and on the twenty-fifth the lifting fog gave them a glimpse of the land, a cliff face, ahead and barely a mile away.

They spent three days of inactivity, yearning for the solidity and moss-covered slopes of the island, but fenced off from it by broken and moving pack with too much water for sledges and too much ice for boats. A projecting cape on their right hand seemed to offer security against their floe being drifted past the land into the open sea beyond. On the twenty-eighth they ferried themselves and their belongings on ice cakes to a floe close in-shore and near fast ice. It proved to be sweeping past the land at three miles an hour. "The southwest cape of the island was not half a mile away, and this was our last chance." The revolving ice cake lodged against the fast ice and a desperate scramble brought all ashore with some immersions but no loss of life or material. They were not even then on the land but on the belt of fast ice which lined its shore—"a confused mass, honeycombed, cracked and broken" impassable to sledges, backed by towering cliffs alive with dovekies. In the rubble at their feet the crew was mustered on the first secure resting place they had known since the ship went down.

The island was named Bennett after the expedition's sponsor, the cape to the southwest, for so many hours anxiously observed was called Emma, after the captain's wife.* De Long determined

* The lonely Bennett Island has other tragic associations than those connected with the voyage of De Long. It was next visited in June, 1902, by the Russian scientists, Toll and Zeberg, and two Siberian natives, who came over from New Siberia with dog team and *baidar* for a summer of field work on the island. They were never seen again. The steamer *Zarya*, appointed to bring them back in September, met heavy ice to the southeast of the island, and after waiting for four days, was obliged to return owing to shortage of coal. In August, 1903, a rescue party crossed over to Bennett Island by whaleboat "in absolute freedom from ice." In a driftwood hut they found scientific instruments, and a dispatch addressed to the Academy of Sciences, St. Petersburg, dated November 8, 1902, and signed by E. Toll. This contained a summary of data obtained in geology, astronomy and wildlife, and added that the four men had pre-

to remain for a week's rest. The crew had endured six weeks of heavy labor and periods of abnormal exertion with extraordinary health and good humor, but the captain was too shrewd to rely on their long continuance. The capture of seal, walrus and bear on the march left him with adequate supplies, and sailing prospects would improve as the month of August advanced. To avert the demoralization apt to result from complete idleness, he assigned light duties: the men to make tidal observation, collect driftwood, eggs and flowers; the officers to hunt game or make scientific observations. Dr. Ambler studied the geological formation of the island. Dunbar followed the south shore around Cape Emma on foot. Mr. Chipp "examined the shore by boat," Mr. Melville discovered coal and tested its quality. With hopes of continuous navigation to North Siberia and on to the continental shore De Long detailed boat crews. He took charge of the first cutter with a total crew of thirteen, Chipp and Melville of the second cutter and whaler respectively, with a crew of ten each. Danenhower still disabled by bad eyesight was assigned to Melville's command to provide a line officer for each boat—a prudent measure to which the crew of the whaler were probably indebted for their survival.

They took their departure from Bennett Island on August 6. Novaya Sibir, the easternmost of the three large islands of the New Siberian group, lay sixty miles to the south. Looser ice now permitted them to take to the boats and thread their way southward through ice lanes, at intervals hauling up boats and cargo and trekking over floes, more than ever rough and shattered. These spells of labor were frequent; on August 9 De Long notes five miles of uninterrupted navigation as "a miraculous piece of good fortune." He now had most of the dogs shot, to the distress of the seaman Ericksen who in two months was to desire the same fate for himself.

---

pared themselves fur clothing "indispensable for a return journey in the winter season." From this it was conjectured that Toll and his comrades had set out on foot for New Siberia and had perished when the newly frozen sea broke up in a November gale.

Stationary old ice was not the only obstacle to the rapid progress of the boats. As the sun receded young ice was appearing with growing frequency on pools and in lanes, and, far worse, gales would lash the ice-strewn sea into a turmoil, and drive the party to take cover in tents on the securest floe they could find. The passage from Bennett Island which might have been covered in a day required two weeks. Only on the afternoon of August 20 did a shadow of land to the southwest assure them that they had completed the first leg of their journey and reached the threshold of known land. For the next few days they drifted helplessly in the pack, in what direction De Long could not at once determine. He took stock of his provisions and found that, though some types of food were exhausted, rations were still in good supply. More immediately trying was the rapid exhaustion of private stocks of tobacco:

> Since the evening of the 22nd [entry for August 26] I confess that I have been perfectly miserable for want of a smoke. This noon, after dinner, Ericksen came to me, saying he had noticed me going around without smoking, and he tendered me a small packet of the precious article, tobacco. I declined more than a pipeful, but he insisted upon my taking more, saying that they had enough for some days in No. 6 tent. I sought out the doctor and Nindemann, and made them as happy as myself.

There was fine comradeship in that doomed band.

When De Long made this entry his party, jammed in the ice and enveloped in snow and fog, had no clear perception of where they were. On August 28 better visibility showed him that the drift had carried the boats further than had been supposed on their course, along the west shore of Novaya Sibir, and pushed them into the channel dividing it from Faddeya Island. On the twenty-ninth a portage took them over congested ice at the narrows of the strait into freer water beyond. On the thirtieth they made a landing on the southwest angle of Faddeya Island. The one remaining dog chased lemmings in a frenzy of delight, while the men gratefully drew water from fresh ponds to replace the brine-tainted liquid in their casks.

From there they went on through storm and shoal water, portaged over ice, jammed against Kotel'nyy Island, and made another landing on its southern shore. Their course now lay southwest to the Lena Delta, not the nearest point on the Siberian shore, but the part offering best chance of the help which they were sure to need. Cape Barkin, where De Long had ordered the boats to rendezvous in event of separation, lay 200 miles west-southwest. This promised to be the speediest and was certain to be the riskiest part of their long perilous journey. The boats had not improved their seaworthiness by being wrenched and dragged over hummocks and let down into leads; consequently the prospect of an open sea on the last leg of the journey was not entirely welcome; ice streams kept the sea down in a high wind and in emergency a large piece afforded refuge from a gale. De Long took the precaution of lightening the undersized second cutter of two of her crew: he took the cook, Ah Sam, into his own first cutter and sent the seaman Manson to Melville in the whaleboat. This proved a lucky escape for Manson, but the unfortunate cook was to profit nothing from this transfer. Before they reembarked hunters were sent out in quest of reindeer and polar bear. They came upon empty huts showing that the coast was sometimes the resort of fisherman, but found no human life and no wildlife but fowl. The friendly help which had saved Borough, Heemskirk and Payer was not to be theirs.

THE *JEANNETTE*: RESCUE MISSION
IN THE LENA DELTA

The boats of the *Jeannette* left Kotel'nyy Island on September 7 and through a sea, open but with ice enough to provide nightly camp sites, made the passage to little Semenovskiy Island in three days. There they halted for a day, repaired the boats and shot a deer.

On the twelfth the three crews took their departure from the island with a fair wind, a stiff northeaster. On their left they passed the islet of Vasil'yevskiy, the last land on their course until they reached the Delta a hundred miles away. The boats clustered together for dinner behind a sheltering ice piece, and late in the afternoon hauled up in the lee of a floe to stop a leak in the whaleboat, the last time that loyal but unlucky band were united. Then they were off again in a strong wind that mounted to a gale as night came on. There was imminent danger of the boats broaching to and capsizing; they were deluged with waves and only by constant bailing kept afloat. The lightweight cutter, less manageable in such boisterous seas, began to fall astern; the whaler was drawing ahead. When Melville shortened sail to correct this, the waves broke over the whaler with a violence that threatened to sink her. De Long beckoned Melville alongside; the engineer shouted that he must give his boat her head or be swamped. The captain made a gesture which he took to signify assent and he shook out a reef and shot ahead. Looking back in the deepening twilight the men in the whaleboat could make out the second cutter, a mere leaf on mighty waters, now floating on a crest, now lost in a trough. As they watched they saw a great wave break over her; she broached to. For a moment she seemed to pause in sharp silhouette with her broadside toward them—a man was discerned standing up against the mast struggling to free the sail, taken aback and jammed against the mast. Then the wave passed under her; she sank into the trough and was seen no more. It was Melville's belief that she foundered that very instant. Chipp, the brave old ice pilot Dunbar, the carpenter Sweetman (the only Englishman aboard and he had well maintained the honor of his race) ended their journey there.

For the moment it was only too probable that others would share their fate. Jack Leach, the whaler's steersman, drenched with every wave, had great difficulty in holding her straight down wind. Melville therefore hauled up on the port tack to steady her, and, not himself a seaman in the strictest sense, gave ear to the sailors who assured him that the boat could not live unless she found shelter behind a stream of ice. Failing that, the only resource was to get out a sea anchor and ride out the gale head-to-wind. With the aid of Danenhower and the boatswain, Jack Cole, Melville used tent poles as frame to a sort of canvas parachute which by dragging in the sea to windward would steady the boat with her canvas-decked bow, and not her unprotected stern, to the oncoming seas. Cole affirmed that the structure was too light to be serviceable.

Danenhower, whose near blindness was no handicap in darkness, took charge of the critical operation of putting the boat about. For a moment she must lie with broadside to wind and sea. Danenhower had observed that the seas ran in threes with a short lull after the third and heaviest wave. He waited for five minutes until he saw the expected band of smoother water coming near and gave the order to go about: "The boat came round, gave a tremendous dive and she was then safe, head to sea." Not altogether. The drag was launched but it was too buoyant and drifted alongside the boat. Cole shouted, "I told yez so." A copper kettle was slid down the connecting line; the anchor sank and took hold. Cowering from the cruel wind and spray, or bailing for dear life when a wave broke over them, the crew of the whaler passed a "sad and dismal" night.

They rode to that anchor for twenty hours. The sun appeared at noon on the thirteenth and the wind began to subside. At 6 P.M. they resumed their course west-southwest. On the morning of the fourteenth they grounded in two feet of water with no land in sight. Sounding with a pole they found a navigable channel to the east, and were following it when threatening weather compelled them to withdraw seaward and pass the night in deep water. On the fifteenth they steered southwest and the next day raised land far to the south, the high hills which backed the flats of the delta. Soon landspits appeared near at hand. Late on September 16 they crept into the mouth of a river and drank greedily; they had been without water for four days. They had entered one of the more easterly of the Lena outlets and were more than thirty miles south of Cape Barkin.

Frost bitten, exhausted and now on short rations, the ship-wrecked crew saw their immediate prospects no better, and hope for the future fading away. The land appeared a wilderness, and the river was broad but so shallow that they could not get within a hundred yards of its bank. Danenhower urged that they should push out to sea again and seek the Cape Barkin rendezvous where De Long had believed (mistakenly) that they would find a settlement large enough to afford them relief. As Melville, full of misgivings, ordered the boat about, he noted despair on the faces of the men and a manifest reluctance to challenge afresh the storms and shoals from which they had barely escaped. There was no mutinous gesture, but the stoker Bartlett observed quietly that the river was as wide as the Mississippi at New Orleans and there must be a passage for the boat if they could but find it. This hint ended Melville's indecision. He cancelled his order and began to grope his way upstream. In the evening they came upon a welcome sign of human life, an empty hut on a bank ten feet above the river, and spent the night within its walls. The warmth of a fire produced severe pains in limbs that for days had been drenched and cramped in an over-loaded boat, and not surprisingly all awoke in worse plight than when they lay down. Maimed as they were they worked up-stream to what they took to be the main outlet of the Lena. There they saw three natives in their canoes and by friendly gestures induced them to come alongside. With pantomime aided by a sketch Melville expressed his wish to be guided to Bulun, the administrative center of the area; the natives intimated that this could not be done. They steered their new acquaintances through a maze of channels down another stream to the Yakut settlement of Jameveloch on the sea shore. The folk there could do little for the strangers but keep them alive: they were crowded into a single hut and fed on decayed fish and gooseflesh. All but three or four of his party were now totally incapacitated, and Melville feared that this diet—the best available—would crown their ills with scurvy and typhoid.

Though badly disabled himself Melville was clamorous to get to Bulun, eighty miles above the head of the delta, but the near-est place where he was sure of contacting Russian officials and getting a dispatch to the United States Department of the Navy. In reply the natives explained that the advance of winter im-peded travel by water, while there was not sufficient ice and snow for travel by sledge.

This was the situation when on October 10 there arrived from the opposite side of Borkhaya Bay Kysmah, a former soldier, condemned for some offense to perpetual exile. He was a quick-witted, intelligent fellow, and Melville, who now had a smatter-ing of Russian, could converse with him with some fluency. He impressed upon the Russian that General Cherniev, governor of Yakutsk province, would punish neglect of the strangers, but handsomely reward all who gave them help. Kysmah repeated the Yakuts' warning that travel was unsafe but undertook to carry a dispatch to Bulun as soon as it was practicable. He set out on the sixteenth, promising (as Melville understood him), to be back in five days. He reappeared after two weeks, bringing letters from the Bulun commandant, and a priest, and also a scrap of paper bearing in English the scrawl,

> Arctic steamer *Jeannette* lost on the 11 June; landed on Siberia 25th September or thereabouts; want assistance to go for the Captain and Doctor and (9) other men
> William F. C. Nindemann,
> Louis P. Noros,
> Seamen, U.S.N.,
> Reply in haste: want food and clothing.

So here was evidence that the first cutter at least had weathered the storm and made land. The two men who had signed the note, Kysmah stated, had been discovered by natives dying of hunger in an abandoned hut at Bulcour near the top of the delta. He himself had met and tried to converse with them; he got the impression that many of their comrades had starved. Melville saw this was a mistake; their note showed thirteen of the fourteen men in the first cutter still to be alive.

Kysmah reported that the commandant of Bulun was on his way with sledges and supplies to bring out Melville and his men, but the engineer in his impatience to confer with Nindemann refused to wait. He set out immediately with one native as escort, missed meeting the commandant on the trackless waste through which his journey lay, and so arrived at Bulun on November 2. He found the two men alone in a wretched little hut. Nindemann was lying down with face averted, too weak and indifferent to turn his head at the stranger's entry. Noros rose and stared at his officer without recognition. Melville extended his hand with a hearty, "Hallo, Noros." The sailor gasped, "My God, Mr. Melville, are you alive?" Nindemann started up exclaiming, "We were sure that the 'whaleboats' were all dead and the 'second cutters' too."

132

Though the two sailors were neglected and in miserable plight, their story afforded little hope that their comrades of the first cutter had been even as fortunate as they. Like the whaler the captain's boat had ridden out the gale of September 12 dragging at a sea anchor, while her crew were deluged with freezing spray. With less patience than Melville, De Long, trying to get under way early on the 13th, when the wind was still high, carried away his sail and again rode to the sea anchor. Late in the evening he proceeded with a jury sail made of a blanket and an old sledge cover.

On the evening of the 12th when Melville hauled up on the port tack, De Long continued to drive before the wind, with the result that their courses diverged. Melville, deflected still further to the left by shoal water, made his landing well down the eastern side of the delta and De Long on its northern face. On the sixteenth the captain's crew sighted a low coast running east and west and, after much fumbling among shoals, were brought to a halt a mile and a half from the land. A raft was fashioned, and some began to freight supplies ashore by wading while others contrived to maneuver the boat to within half a mile of land. The water was filmed with ice; the waders, too weak to raise their feet and crush it down, shuffled ahead in anemic fashion brushing it away with their knees. By the evening of the seventeenth all were ashore, with some rations in hand, and noting the welcome appearance of driftwood along the beach. The aspect of the crew was less encouraging: the men whom the toil of the summer journey had failed to subdue, who had reached New Siberia in the best of cheer, were now drenched, cramped, frostbitten and badly dispirited. "That terrible week in the boat has done us a great injury," noted De Long. Sagging morale was further depressed by the absence of any sign of human habitation. Which way were they to turn? The map of the delta which De Long carried showing water courses and places of settlement was an untrustworthy guide. The natives were migratory. The spring melt which flooded the delta was prone to obliterate old channels and carve out fresh ones, making even a well-drawn map as obsolete as an old almanac: De Long's map showed eight mouths to the Lena, the searchers for Chipp and his crew later declared that it had two hundred.

Unaware that not far to the west there lay a village which could have supplied his immediate wants, De Long saw no other course than to follow the shore to the nearest river mouth and push up stream for the permanent settlement of Ku Mark Sarka, which lay ninety-five miles up the delta on the near side of Bulun. It was an all but hopeless endeavor, unless—as there was good reason to hope—they found help on the way.

They set out on September 20 over the frozen surface of a spongy morass, in some places hard and slippery, in others so thinly filmed with ice that the enfeebled wayfarers broke through and plunged knee-deep in mud and swamp grass. But they were encouraged by finding traces of deer and soon arrived at a river coming down from the southwest and wooded on its farther side. On its frozen margin they found better walking as they turned inland and left the sea behind.

But already they were experiencing a handicap that was inevitable in so large a party after months of toil and exposure. Three men partly disabled were retarding the pace of their vigorous and still hopeful comrades. Ericksen whose feet had been badly frozen in the boat begged them to leave him behind to die. De Long rated him and drove him on. From then on he or Dr. Ambler kept near Ericksen to keep him in line. The two days' rations in hand were enlarged by the killing of two deer. But the ill luck which was to haunt their countryman Greely in Kane Basin was theirs also. On the twenty-eighth the sluggish waters of the river were seen to divide ahead, imprisoning them in a fork between two streams, neither of them fordable nor yet firmly frozen. Driftwood was found but no lashings to bind it into a raft. They waited for three days until the river was frozen solidly—the doctor in the meantime ministering to Ericksen by scraping the dead flesh from his toes. They resumed the march finding no game but a few ptarmigan: on October 4 De Long noted four pounds of pemmican remaining and a half-starved dog, for fourteen men. The sight of empty huts and neglected fox traps was disheartening because they suggested a general exodus of inhabitants from the lower delta in the winter season. On the fifth and sixth they took shelter from a gale in a hut and there Ericksen died. His body was sunk through a hole in the ice and a board set up in his memory on the river bank. The journey was nearly over for all but two of the rest. On the seventh a gain of three miles was registered, apparently on the strength derived from the soup of one ptarmigan. De Long's entry for the 8th reads in part:

> October 8th. . . . Went ahead until 10.30; one ounce alcohol 6.30 to 10.30; five miles; struck big river; 11.30 ahead again; sand bank. Meet small river. Have to turn back. Halt at five. Only made advance of one more mile. Hard luck. Snow; S.S.E. wind. Cold camp; but little wood; one half ounce of alcohol.

No food, but alcohol to drug the raging of the stomach. Supposing that Ku Mark Sarka was only twelve miles ahead (actually it was four times that distance), De Long detailed the seamen Nindemann and Noros to go forward for help, furnishing them with blankets, rifles, ammunition and two ounces of alcohol.

The two men set out on the ninth, bridging weak ice at the center of a stream with driftwood, killing a ptarmigan, and observing drawn up on the river bank the flatboat which was later to be a pointer to the spot where the bodies of their comrades lay. On the night of the tenth they used their sheath knives to carve a dugout in a drift under the bank, alternately sleeping for periods of five minutes and exercising to keep circulation alive. On the thirteenth two crosses were observed at the foot of a bank fifty feet high. They climbed this and found an old hut with two boxes nearby. One of these contained a frozen corpse, but there was fish in the hut; this together with boiled boot soles, a piece of sealskin trousers and a lemming washed down with a distillation of arctic willow gave them strength to drag themselves on for six days more.

From time to time huts were found containing scraps of rotten fish; on the nineteenth they waited in the vain hope that an owl flying in circles overhead would alight within range. That afternoon their course led them over a frozen stream. Looking back they saw three huts on the crown of the river bank behind them. It was the spot which Melville was to mark on his chart as Bulcour. The huts proved to contain fish, but so rotten that it gave the starving men violent dysentery. They rested there until the twenty-first.

About noon on the twenty-first a sound was heard outside. Nindemann, thinking that it was a reindeer, seized the rifle and flung open the door to confront a startled Yakut. The poor fellow, mistaking the meaning of the firearm, began to plead for mercy. The sailors quickly reassured him and tried to make him understand their story and the plight of their shipmates. He replied by signs that he must go and promised by emphatic gestures to return, whether in so many hours or days the anxious strangers could not decide.

Their weakness left them no choice but to await his return and hope for the best. In the meantime lacking the strength to gather wood they kept their fire alive by breaking up a sledge and dismantling the sleeping berths in the hut. In the evening their friend came back with two companions and sleighs, fed them fish, then wrapped them in deerskin and carried them an estimated fifteen miles into the hills which overlooked the lowlands of the delta. There in a mountain ravine were two tents in which the men were lodged separately. A woman offered Nindemann water and observing that his hands were bent and rigid and his nails like claws, tucked up her sleeves and scrubbed his face for him.

Both then and on the following morning the sailors made frantic endeavors to acquaint the natives with the plight of their comrades and their desperate need of assistance. The latter responded only with gestures of sympathy. They loaded their sleighs with deer meat and in a body descended from the hills to the river's main channel and in two days carried the seamen upstream to a cluster of huts, the Ku Mark Sarka which De Long was never to see.

The villagers eyed the two strangers with curiosity, but in the season of autumn feasting did not pay serious attention to their clamorous appeals. The unhappy men sat down and sobbed in despair and frustration. A woman, pitying them, took them to a man who said something about the commandant, "komanda." Nindemann in reply implored him to take them to Bulun. That evening the villagers brought to them a Russian whom from his bearing and air of importance they took to be an official. This was no other than the banished Kysmah on his way to Bulun with Melville's dispatches. He did not understand Nindemann's German, but in response to his gestures and fragments of Russian pronounced the words "Jeannette" and "Amerikansk" with an air of intelligence. The sailors were puzzled to note that while he appeared to understand their story of disaster, he seemed wholly unsympathetic to their pleas for help. Only later did they know that he supposed that they were stragglers from Melville's party and that those for whom they sought aid were already safe. Nindemann could do no more than hand him the note which was to inform Melville of their survival and passively accompany a native escort on the road for Bulun.

A dramatist's imagination could never create a situation of more tragic pathos than that of Nindemann and Noros at Ku Mark Sarka. Their comrades lay perhaps not more than fifty miles away; the strongest of them still had a week to live. They died because their loyal shipmates, after enduring frightful toil and privation on their behalf, lacked the speech to make their condition plain to those who were able and willing to help.

The Bulun commandant, Brieshoff, received the sailors kindly, forwarded to the American ambassador at St. Petersburg a dispatch composed by Nindemann, and declared his intention of going to the "captain," meaning not De Long, whose situation he may not have understood, but Melville on whom Kysmah had given a report. He set off the next day in such haste that he neglected to give specific orders for the proper treatment of Nindemann and Noros. The people charged with their care suspected that they were impostors and runaway convicts whom it would be dangerous to treat with liberality, and so housed and fed them wretchedly. The local priest, less in awe of the civil power, showed them what kindness he could and eked out their stale fish with black bread. In this condition Melville found them.

On hearing their story Melville regretted taking the precipitate journey which had actually deferred his meeting with Commandant Brieshoff. His first care was to secure proper treatment for his neglected comrades. Not relying on his appearance and authoritative bearing which impressed natives and European Russians alike, he went himself to break open the door of a decent hut and saw to it that his men were well housed and fed before procuring a fresh escort and starting again downstream.

At Burulak, between Bulun and Ku Mark Sarka, he met Brieshoff's caravan of sledges bearing Danenhower and the rest of the whaleboat crew, all in health save Jack Cole, once the mainstay of the party, but bereft of his wits since the disastrous night of September 12. The commandant, "a fine specimen of Cossack manhood, very large of stature, of a commanding presence and quiet demeanor," received Melville cordially, promising to send him and his party a thousand miles up the river to the provincial capital of Yakutsk. Melville thanked him on behalf of the rest: for himself he requested dogs and drivers for the search which, in spite of fading hopes and the rapid advance of winter, he was determined to make for De Long. He was given two Yakut drivers and with them descended again to the bleak and windswept delta.

There can be few instances of heroic and unselfish endurance grander than that displayed by Melville for the three weeks of his search. Winter temperatures on the Lena Delta are not so low as in some parts of Siberia owing to the nearness of the ocean, but it is reputed the worst terrain for winter travel owing to the winds which sweep without check over its alluvial surface. Melville's only companions were unsympathetic half-barbarians whose language he imperfectly understood; he had not recovered from the hardships of the autumn voyage; his feet and legs were swollen to nearly double their natural size and covered with water blisters which the sympathetic Yakut women who gave him shelter would dry and smear with goose-grease. Yet he persevered until the grudging twilight of winter gave way to complete darkness.

His first objective was the hut at Bulcour where Nindemann and Noros had been found. He wished to retrace the course they had described to him to the place where they had parted from De Long, being assured that the captain's party were too weak to have advanced far from the point of separation. On the way down his guides mutinied, left part of their rations behind and professed their inability to go farther. One of them feigned starvation and spoilt the act by simulating a wolf. Melville thrashed one of them, *pour encourager l'autre*, and assured both that he would eat the dogs and then them before quitting his search. Finding him so unyielding his guides went on with docility, directing him from time to time to settlements where food and healing warmth could be found.

At one of these resting places a native gave him a record planted by De Long on his upland journey. Melville perceived that he had strayed from Nindemann's track and overshot his mark. Rather than make a random search in that maze of frozen waterways, he went down to the ocean shore and found a point of reference in the flagstaff marking the first cutter's landing place. He then went upstream, guided by Nindemann's recollections and keeping a lookout for the monument to Ericksen which would mark a near approach to his objective. He missed the trail again, and a furious blizzard drove the travelers to the shelter of a cavern hollowed out of a drift. Suffering from frostbite, and too weak to walk, Melville bowed to the inevitable and headed south for Bulun which he reached on November 27. He had achieved no concrete success, but had simplified materially the next search, which he was determined to make in late winter before the spring floods came to blot out all traces of De Long's catastrophe.

At Bulun Melville met Epachiev, assistant *espravnik*, or governor, "a Russian, born in Yakutsk, sociable, intelligent, and withal a very fine fellow," with whom, despite frequent misapprehensions, he was able to discuss religion, politics and geography. With this man and those of his comrades who had not gone forward with Danenhower he set out via Verkhoyansk to Yakutsk. With Kashorovski, espravnik of Verkhoyansk, Melville was able to converse freely through a student interpreter who spoke English. This young man was one of five exiles banished to Siberia as nihilists. Melville asked permission to visit these men, which the espravnik granted with the indulgent remark that the conversation of a nihilist, however damaging to a loyal subject of the Russian monarchy, could do little harm to a republican from America. The five exiles, ranging in age from seventeen to twenty-nine, were all devoted revolutionaries, though some declared that they had not been so until exiled for no known reason. Melville was amused at their belief that the Communist system to which they were dedicated was already established in the United States. They plied him with questions as to how he arrived in Siberia and did not conceal their hope of obtaining a boat and making their escape by way of the River Yana and the sea. Melville learned subsequently that they did make this crazy attempt and in consequence were banished to a station even more remote than the dismal outpost of Verkhoyansk.

Whatever the Americans thought of the Russian system of government they had only one opinion of the kindness to themselves of those by whom the system was administered. General Cherniev, governor of Yakutsk province, sent Danenhower five hundred rubles from his private purse to meet expenses on the journey from Bulun to Yakutsk. On arrival he lodged him and his men comfortably, arraying the lieutenant in "tight boots, white shirts, and choker collars," clothing suited neither to his previous habits nor his present environment. On learning of Melville's arrival at Yakutsk, the general in a neat uniform which fitted him perfectly, "straight as a spear-shaft, and rather spare, with full flowing white hair and beard," left his cabinet and met him on the threshold, repeatedly embraced the fur-clad, frost-scarred veteran, and called him brother. Melville and Danenhower dined with him daily throughout their stay in Yakutsk.

This period was not prolonged. Melville's dispatch, forwarded from Bulun to the nearest telegraph station, brought a prompt response from the secretary of the Navy. The chief engineer was authorized to remain in Siberia to make search for the missing crew and to incur any expense which might forward his purpose. General Cherniev declared that with this authorization Melville had the whole Russian empire at his back. Bartlett and Nindemann had already offered to share in the work; others of their comrades stifled the yearning for home and volunteered to join in with them. Danenhower and the least fit of the crew members left for the United States via European Russia. Melville's constitution appears to have remained unimpaired by six months of extraordinary exertion and hardship. He reached Yakutsk on 30 December 1881. On January 27 he started north again with five men—three interpreters, one an exiled officer, Bartlett, and Nindemann.

Stopping at Verkhoyansk he learned that his acquaintance Kashorovski had been reposted to a lonely station near the mouth of the Kolyma. When he protested at this apparent demotion General Cherniev replied that "he loved and trusted him, and consequently could send him a long way off, while it was necessary to keep the rogues nearer home." Melville added his valued friend Epachiev to the party and continued on to Bulun.

Under the direction of Russian officials natives were employed to set up food depots which would permit two search parties to operate—one at Mat Vai twenty miles above the point where Nindemann had left the main party—the other at Kas Karta downstream. The subsequent revelation that Kas Karta lay on quite a different stream from the one followed by De Long, and perhaps thirty miles west of his actual route, illustrates the difficulty of Melville's task in a labyrinth of rivers and low banks, different but hardly distinguishable. While the depots were being prepared, he journeyed to Jameveloch to reward the natives for their services of the previous autumn. Word had got abroad among the Yakuts that the Americans had unlimited funds, and caused those semibarbarians to adopt the civilized practice of forming an association to rig prices. Melville, the Scot, complained to the espravnik who used his authority to peg prices at an equitable level.

It was Melville's intention to take up the search where he had left it in the autumn and work upstream, using the hut where Ericksen had died and the flatboat which Nindemann had seen as pointers to De Long's last resting place. On March 18 he and the quartermaster mustered native drivers and dog teams at Kas Karta. Nindemann declared that the first cutter party had not passed that way. As the river divided there in its downward flow they followed the eastern branch and arrived at a junction of streams called by the natives Usterda and recognized by Nindemann as the place where they had crossed the ice in a westerly direction on October 1. But again the thread was lost. Nindemann could not be positive in his identification of landmarks now almost buried in snow and Ericksen's hut was not to be found. So they went up twenty miles to where the Lena paused in a sort of lake or reservoir and sought a clue in one of the outlets through which it spewed its waters over the delta below.

At the top of a bluff bank they found the remains of a large fire. It was Nindemann's belief that he and Noros had passed that way shortly after parting from De Long. To confirm this he drove down the river to look for the flatboat on the beach. Melville was following him when he observed at the top of the ramp of snow sloping down from a cut bank four stakes lashed together and supporting a Remington rifle. Supposing this to be a marker for cached records he trudged up the incline to fix the spot with compass bearings. On the crest, windswept and almost bare, he saw a teakettle and then came upon the remains of a fire and the bodies of three men grouped around it, those of De Long, Dr. Ambler and Ah Sam. De Long lay on his right side with his face to the fire and his left arm extended upward. Behind him lay the journal in which the last days of the journey were recorded. Melville believed that he had died in the very act of tossing it backward away from the fire.

De Long's daily entries were kept up to the last. After the parting from Nindemann and Noros on October 9, their comrades toiled on, supported by nothing but a little alcohol, glycerine, and scraps of deerskin, eked out with a distillation of the Arctic willow. Alexey, the Indian, died on the eleventh. His body was laid in the abandoned flatboat before being carried on to the river and covered with ice. His comrades after advancing twelve miles in eleven days made their last camp on October 20, sheltered by a recess in the river bank. Two men died on the twenty-first, and as the seamen were too weak and indifferent to stir, the captain, the surgeon and Mr. Collins carried their bodies out of sight around an angle in the bank. The others passed away one by one where they lay. De Long's last entry, for October 30, read: "One hundred and fortieth day [since the loss of the ship], Boyd and Gortz died during night. Mr. Collins dying."

Evidently the three men who still lived, De Long, Ambler and Ah Sam, had mustered up the strength to carry the chart case up to the top of the bank, intending to do the same with all records and save them from the spring floods. Finding this beyond their strength, they had kindled a fire, brewed tea from the arctic willow and remained to die on the spot. From De Long's record Melville discovered that the bodies of seven men, all those unaccounted for except Alexey, lay five hundred yards away, in the gulch where their last camp had been set up. These were quarried out of the snow and carried up the bank. The hardened drift under the river bank which covered the bodies would not have melted until the floods came to wash it away; and but for the resolution of the three weak, numbed and dying men, who clambered out of the riverbed to die on the crest, their exact fate would never have been known. Since his Yakut guides assured him that the whole region would be ingulfed in the spring inundation, Melville carried the bodies into the foothills to the west and buried them under a pile of stones surmounted by a large cross. They were later disinterred and taken home for burial.

It was now early April. Melville and his two principal helpers had endured ten months of abnormal hardship and tension. Yet they would not rest until they had done what they could to ascertain the fate of Lieutenant Chipp. The second cutter had not been *seen* to sink. There remained a faint possibility that she had come to land. Melville divided his force into three parties and by sections they examined the entire seaward face of the delta and extended their search eastward to the mouth of the river Yana. Neither they nor Lieutenant Harber of the United States Navy who continued the quest in that and the ensuing summer uncovered the least evidence that the second cutter or any of her crew had reached the Siberian shore.

While the chief engineer was thus engaged he heard of the presence in eastern Siberia of the celebrated American soldier, traveler and journalist, H. G. H. Gilder. In 1881 this unwearied adventurer came back from the Canadian Arctic where he had been retracing with Lieutenant Schwatka the death march of Franklin's crews. On learning that the U.S.S. *Rodgers* was fitting out a similar mission on behalf of De Long, he secured the post of pay clerk and shipped as one of her crew. When his ship was burned at her winter anchorage in the Bering Strait he set out for the Lena by land and on the Kolyma met Kashorovski who told him of Melville's successful search. Later on Gilder met the courier bearing Melville's dispatches relating to the discovery together with extracts from De Long's journal. Gilder induced the luckless Cossack to open his packet, transcribed such parts of the dispatch as interested him, and forwarded the fruits of this felonious "scoop" to the New York *Herald*. The indebtedness of the expedition to the proprietor of the *Herald* deterred Melville from any harsher censure than a caustic reference to the "spirited journalist," but he must have been not a little mortified by the severe remark of General Cherniev that such conduct might be permissible in free America, but that in Russia it was an offense of the gravest description. One wishes that Melville had interceded to shield the unfortunate Cossack courier from punishment (he had probably regarded the hustling American as a high official and dared not refuse him). Doubtless, however, the old martinet of a general would have held that such an offense as tampering with the mails called for severe retribution regardless of the offender's intent.

On his way back from the Yana to Yakutsk Melville had proof that the natives had not exaggerated the extent of spring inundations. He was held up by floods and quoted a report that Gilder and others had suffered the humiliation of being driven up a tree and had lived for days off the carcase of a horse moored to its trunk. Melville arrived at Yakutsk almost exactly a year after the loss of the ship and returned home by way of Europe.

A naval court of inquiry appointed to investigate the loss of the *Jeannette* found that the ship was fit to polar service according to the standards of the time. De Long, Melville, Nindemann and the carpenter Sweetman were specially noted in a commendation passed on the conduct of the entire crew. The latter was more than deserved: it is doubtful whether in that century of bold adventure, uncalculated risk and not infrequent disaster any party of men fought harder or with better courage and discipline for their lives.

THE NORWEGIANS: THE *FRAM* ON THE
NORTH POLAR CAP

The brave but unfortunate De Long died without the satisfaction of knowing that his enterprise, which appeared to have terminated in utter disaster, was to be the prelude to a deeper and more prolonged penetration of the frozen Arctic than had yet been achieved.

The *Jeannette* had sunk on 12 June 1881 in 77° 15′ north latitude and 154° 59′ east longitude. In the autumn of 1884 it became known in Europe that a number of articles belonging to her had been found by Eskimos in the ice off southwest Greenland and delivered to Mr. Lytzen, colonial manager at Julianhaab. Among these were a list of provisions, signed by De Long, a list of the *Jeannette*'s boats, a pair of oilskin breeches marked "Louis Noros," and the peak of a cap bearing the name of F. C. Nindemann. The character both of the men who made the discoveries and of him who reported it made forgery improbable, and the fact was readily accepted as genuine and of the greatest importance. Supposing the relics to have traveled in a more or less straight line from the scene of the wreck to the south Greenland shore, they had drifted thousands of miles on a course which took them close to, if not right over, the summit of the globe within three years. The inference was that, had De Long's ship been stronger and supplied for the requisite period, he might have drifted to the North Pole and made a safe return by way of the east Greenland shore and the North Atlantic. Though there were not wanting men at the time to draw this inference, some years elapsed before one appeared with the courage, ability and financial backing to give it practical application. Great Britain and the United States, long foremost in the drive to the north, were disheartened by the costly failure of Nares, and the catastrophes of De Long and Greely, and it was to be years before the tenacious Peary won popular notice and support. The man who appeared to exploit the De Long find was the Norwegian, Fridtjof Nansen.

This celebrated individual belonged more naturally to the spacious Elizabethan era than to the blasé and decaying nineteenth century. In the diversity of his gifts and interests he resembles the great Raleigh, to whom, though perhaps inferior in the gift of imagination, he was vastly superior in steadiness and moral principle. Like Raleigh he has left posterity rather puzzled as to where his chief excellence lay. An historian, a scientist in more fields than one, in the front rank of polar travelers, patriot, and a diplomat, who after the age of sixty as the agent of the League of Nations worked for the repatriation of prisoners of war and starving Russians and Armenians, and having fairly earned the title of first citizen of Europe, Nansen died beloved by all clashing ideologies.

Nansen was twenty-nine when he first propounded the scheme of a deliberate transpolar drift. Son of a Christiana (Oslo) lawyer, he was curator of the Bergen Museum at twenty-one; in 1888 he produced an original study of the nervous system. It was possibly Peary's ascent of the Greenland ice cap in 1886 which diverted his ambition from purely scientific achievement to northern travel. He organized and led the Norwegian skiing expedition which in 1888 crossed the Greenland ice cap from Umvik fiord on the eastern shore to Godthaab on the west. Thus he won the prestige necessary for the promotion of the far more costly and dangerous transpolar plan.

In February 1890 Nansen outlined his proposals before the Geographical Society of Christiana. These were: to build a ship light and so shaped as to slide upwards in ice pressure; and to put her in the ice at the New Siberian Islands, near the site of the *Jeannette's* shipwreck, in the hope of drifting west, as the relics of the *Jeannette* had done, perhaps right over the Pole, or at any rate, deep into the unexplored area which lay around it. Comfortably housed in a ship, the explorers would be infinitely better equipped than sledge parties for scientific work and would be spared the hardships hitherto inseparable from polar journeys. Years of comparative inactivity would pose a serious threat of scurvy, but Nansen expressed his conviction that by proper diet and regular exercise that particular danger of the polar climate could be overcome.

In his own country Nansen's proposed venture found generous and ungrudging support. The Norwegian Storthing voted a sum equal to two-thirds of the estimated cost. King Oscar, the ruler of Norway and Sweden, made a handsome donation; a host of contributors, native and foreign, made up the required total. Among the latter was the Royal Geographical Society of Great Britain; and before that body Nansen appeared to expound his theory and the method by which he proposed to apply it. The individual members of the Society were far from convinced of the soundness of the project to which, as a body, they had contributed. Sir Leopold McClintock, whom Nansen particularly admired, admitted the probable soundness of his theory, but shrank with a sailor's instinct from the scheme of deliberately putting a ship of whatever shape and strength into an area of prolonged and incalculable pressures. Sir George Nares denied the validity of the drift theory and, like a methodical naval officer, protested against the impropriety of exposing personnel to danger without an ascertained line of retreat. On the other side of the Atlantic Greely, the best known American with Arctic experience, expressed himself in terms of undisguised contempt.

Nansen does not tell us how he was affected by these adverse opinions. In any case he was fully committed. Nor was he without encouragement. Sir Edward Inglefield, who forty years before had shown the Americans the way into Kane Basin, applauded the scheme, as did the geographer, Sir Clements Markham, a circumstance which ought to be remembered to his credit, as that illustrious Briton was prone to be ungenerous to any project "not British" in origin.

The ship on which Nansen founded his hopes, the *Fram,* was launched in the autumn of 1892. Equipped with both sail and steam, she was the first vessel to be especially constructed for Arctic navigation. Without keel, broad of beam and round-bottomed, she had the semioval shape of the modern icebreaker. Naturally she was not expected to be a comfortable boat in rough weather or good at sailing into the wind. She weighed 400 tons and was to carry a crew of thirteen supplied for five years.

As captain, Nansen secured the services of a comrade of his Greenland journey, the skilled navigator, Otto Sverdrup. (This man was later to become a great discoverer in his own right.) Scott-Hansen, a lieutenant in the Norwegian navy, ranked next to Sverdrup among the professional seamen; for the rest Nansen had no trouble in manning the *Fram* with crewmen who could double up in some scientific or technical capacity. A Reserve Army lieutenant, Frederik Hjalmar Johansen, rather than forfeit a berth on the *Fram,* signed on as a stoker.

The *Fram* sailed from Christiana on 27 June, 1893, circled Norway to the north, calling at various ports en route and took final leave of her homeland from Vardo. The passage of the Barents Sea, where fog and ice gave her crew a foretaste of what lay ahead, proved the soundness of the ship's design. She passed under Novaya Zemlya by way of the Yugor Strait, where English and Dutch had been brought to a halt three centuries before, and put in at the village of Khaborova to embark the dogs which the merchant Trontheim had collected from the Russian interior. On the night of August 3–4 the *Fram* set forth again out onto the waters of the Kara Sea.

In Nansen's judgment the most critical part of the three years' cruise lay immediately ahead. He had few doubts about the soundness of his plan or the fitness of his ship for its execution— once she was imbedded in the New Siberian ice pack. He had less confidence in reaching that station in the current season. His chief obstacle was likely to be the sea off Cape Chelyuskin, nine hundred miles to the northeast; but, the seas between him and it were studded with shoals and islets, inaccurately laid down (where they were laid down at all). In these waters where caution and deliberation were most required, the Norwegians were in a desperate hurry. Halted by ice off the Yamal Peninsula they landed to hunt reindeer. On the ninth they were again at sea, threshing into an east wind against which the broad and keelless *Fram* made poor headway. She had failed to meet her supply ship at Khaborova and so coal for the engine was in short supply. An attempt to set a course straight for Cape Chelyuskin was thwarted by ice: there was no choice but to crawl along the Siberian shore. On August 22–24 they were held up by a wind too high for maneuvering the ship or for hoisting out a boat to find a safe channel by sounding. A snowstorm brought the warning, easily comprehended, that continued delay might convert the temporary anchorage to a permanent one. On the twenty-fourth the wind subsided enough for the low-powered *Fram* to get clear of the shoals under steam and she continued her course among islets where Nordenskiold's observations made in poor visibility were of that vagueness which disturbed the navigating officer instead of giving him comfort and assurance. Much time was lost in finding a way through the intricate island pattern in Taymyr Bay. September 5 found the ship hove to east of the Taymyr Peninsula, with Nansen forming plans for explorations, if, as now appeared all too probable, they must winter where they were.

September 6 was the commander's wedding day and his hopes of a lucky break on that anniversary were realized. As the gale which had held them prisoners blew itself out it was seen to have swept the channel ahead clear of ice. On September 6–7 they worked their way up the west side of the peninsula between land and ice pack until they were stopped again by ice jammed against the land. On the morning of the ninth a bright sun revealed that heavy squalls had loosened the obstructive pack. Sverdrup had scruples about taking the ship up a restricted channel beset by ice and perhaps by shoals; Nansen, impatient and on the verge of success, overruled him. So up they went and soon raised far to the north the mountains which backed Cape Chelyuskin. Late on the evening of the ninth the cape itself was plainly in view under a solitary star; at sunrise on the tenth they were rounding it, a low cape backed by a flat rocky escarpment, its face seamed with snow-filled gorges. Punch, fruit and cigars were distributed to all hands to celebrate the attainment of the northernmost continental land on the globe.

From Cape Chelyuskin to the intended station off New Siberia the direct voyage was short; but ice compelled the *Fram* to creep down the east side of the peninsula and stay by the land until near the mouth of the Anabara, where the edge of the permanent ice began to recede northward under the influence of the warm water disgorged from the Lena's multiple outlets. Setting a course a little to the north of west the *Fram* steered straight for the New Siberian Islands until September 18, when a little to the north of the Semenovski Island where De Long had mustered his boat crews for the last time, she was able to wheel to the left and strike north in warm air and over an open sea. Nansen was exultant. He wrote on September 18, "It was a strange feeling to be sailing north in the dark night to unknown lands, over an open, rolling sea, where no ship, no boat had been before." On September 19, "I have never seen such a splendid sail. On to the north, steadily north, with a good wind, as fast as steam and sail can take us. . . . One must have gone against the stream to know what it means to go with the stream." But, on the twentieth, "I had a rough awakening from my dream." At 11 A.M. the ship luffed suddenly to avoid running on to "the edge of the ice, long and compact, shining through the fog."

She was in 77° 44′ north latitude. She coasted westward along the ice edge but soon became entangled in ice and on the twenty-fifth was permanently frozen in, 300 miles to the west and a little to the north of the spot where the *Jeannette* had gone down.

In the eighteen months which elapsed between 25 September 1893 when the *Fram* was frozen in some 200 miles north of the New Siberian group, to 14 March 1895, when Nansen and Johansen quitted the ship to attain on foot a higher latitude than the drift had been able to supply, the ship was carried, with some setbacks and deviations, in the general direction which Nansen had forecast, from about 79° north, 131° east, to 84° 10′ north, 101° 55′ east, some 500 miles. Nansen's conviction that proper food, exercise, routine work, literature and other entertainments would preserve health and morale was amply justified. There was no sickness—Dr. Blessing in despair gave up the crew and took to examining and treating the dogs. Electric light from power generated by a windmill provided a luxury hitherto unknown in the high Arctic. Nor were the gloomy predictions of British experts that the *Fram* must sooner or later be caught in a nip borne out. Several times she was threatened by an advancing pressure ridge, when a moving ice field began to crush the opposing sheet and roll it back on the ship in a mounting heap of rubble. Had the ship, as McClintock feared, remained fixed she would have been crushed, but she never failed to mount up buoyantly, nestled in the shattered ice that had threatened her, with no graver consequence than the rubble which spilled inconveniently over the bulwarks onto the deck. About the time that the ship was frozen in the crew were gravely annoyed to find that they were sharing their quarters with a colony of bedbugs. When civilized methods of ridding themselves of these pests proved ineffective, they bundled up all infested clothing and bedding and left it on deck for the term of the Arctic winter—a remorseless expedient, but one that was crowned with success.

Nonetheless Nansen was not altogether spared the anxieties

certain to trouble the adventurer who is depending on forces imperfectly understood and which he is powerless to modify or direct. At the outset he was given cause to doubt the basic theory of the cruise; the drift was carrying the *Fram* in the direction opposite to that desired; on November 19 she was five degrees *east* of her starting point. Nansen wondered if he was to return as he had come, by way of Cape Chelyuskin. A month later the ship had been pushed up beyond 79° north latitude, and her commander's confidence was in a measure restored. He had no other grounds for anxiety. Seal, reindeer and walrus had been killed on the way up, and here the polar bear, numerous even in that remote and frozen sea, furnished both sport and an additional insurance against scurvy. Christmas found all hands relaxed and even growing fat: "I am almost ashamed of the life we lead with none of those darkly painted sufferings of the Arctic night."

Though Nansen plainly hankered after the glory of actually reaching the geographical Pole, he constantly disclaimed this ambition in his diary—the purpose of the voyage was to explore within the inner circle of the polar cap—though, of course, the deeper the penetration the better. He was not long denied the gratification of a discovery of great significance. Hitherto it had been supposed that the inner Arctic was a shallow and perhaps island-studded sea, like the Canadian archipelago. This soon proved to be an error. By 4 January 1894, the drift, though disappointingly slow, had carried the *Fram* over the continental shelf into water where no bottom was found in 1,000 fathoms—over a mile in depth. For the moment this discovery was cause for distress. In the deep ocean there was less likelihood of finding the powerful, unvarying current on which Nansen's plan was founded, and his conscience smote him for having impetuously led his men into danger on an unproved hypothesis. "This at once upsets all faith in the operation of a current; we find either none, or an extremely slight one. Columbus discovered America by means of a mistaken calculation, and even that not his own; heaven only knows where my mistake will lead us." In his despondency he was disposed to question the genuineness of the *Jeannette's* relics, but added: "Only I repeat once more—the Siberian driftwood on the coast of Greenland cannot lie; and the way it went we must go."

That way they went though with exasperating slowness, borne on a current which carried them gradually and with setbacks to the north and west. On 2 February 1894 they crossed latitude 80° north in longitude 132° 10′ east, only to be carried back thirteen miles a week later. "It is extraordinary how little inured one gets to disappointment." Progress was even slower in spring and summer of 1894; from March 8 to September 5 the *Fram* traveled barely sixty miles to the west with almost no gain in northing. With all his courage Nansen was liable to fits of depression which he confided to his journal alone. In mid-June he voices uneasiness as to his situation in the heart of an irregular wilderness of decaying ice which resembles a huge city overthrown by an earthquake, filled with mounds of rubble so pulverized by pressure and rotted by the sun as to be barely passable even on snowshoe. In Norway Nansen had confidently reckoned on reaching land on foot and by boat if the ship were crushed; but he was now oppressed by a real fear of disaster: "The ice gets softer and softer every day, and large pools of water are formed on the floes all around us. In short, the surface is abominable. The snowshoes break through into the water everywhere. Truly one would not be able to get far in a day now should one be obliged to set off towards the south or west." Taymyr Peninsula, he recalled, was totally uninhabited, and he was far further from the nearest settlements than De Long had been.

In the midst of these anxieties he rejoiced (as in a like situation McClintock had done) that his men were free of the burden that weighed so heavily on their chief. Cheered by the eternal sunlight and by the abundance of bird life—gulls and kittiwakes—

they passed the summer in cheer and good health. Nansen himself found relief in the treasures of marine biology brought up by the dredge.

On September 2, the first day of autumn, he was able to record a net gain of 189 miles in the course of the year, and in a more hopeful vein reaffirmed his confidence in completing the transpolar drift in three years. With the advance of winter movement seemed to accelerate: On October 21 the men of the *Fram* were able to celebrate the crossing of the 82° north latitude, and on December 13 they crossed the 82° 30′ line, so breaking the record of Sir George Nares and setting a new record for the highest north made on ship.

On 3–6 January 1895 a disturbance in the ice subjected the *Fram* to the sharpest test of the whole cruise, a test, Nansen declares, which not another ship in the world could have withstood. A floe, moving in on the ship's port side, overrode the ice in which the *Fram* was embedded and bore down until the ship herself was heeled over several degrees. Water came welling over the ice and flooded the kennels of the dogs, driving them to take refuge on the sloping deck. Coming on with intermittent and barely perceptible motion and pushing before it a mounting heap of rubble the floe took the *Fram* amidships burying her port rail in ice slag while the vessel within groaned under pressure. All doors below deck were opened to prevent them from being jammed shut in the squeeze; supplies were packed, dropped over the starboard rail and stowed on a stout old floe nearby. Loaded sledges stood on deck ready for instant departure; the mate secured to his clothing cups, knives and other small articles for which no other means of transport could be found. "In marching order on an empty ship," the men, quite in the Viking tradition, fell to devouring what they could of those delicacies which might soon be left behind. Too anxious, one guesses, to take part in these revels, Nansen noted that the ship had partially righted and he conjectured that she had jumped loose from her icy bed and could be trusted to rise if pressure were renewed.

The next day she was found to have risen a foot forward, six inches aft beside slipping astern a little. At the same time a change in the wind put an end to pressure at the point of immediate danger, the port side, and notwithstanding Nansen's worried consciousness that they were "living on a smoking volcano," the feeling of crisis began to fade, and thoughts to be diverted to other matters. A meridian observation by Lieutenant Hansen showed that they were in latitude 83° 34′, so breaking the highest north record set by the Americans Lockwood and Brainard on the north Greenland shore. Living man had never been as near the pole as they.

Long before the achievement of this record Nansen had formed a plan which, though received by his crew with comparative nonchalance, seems the boldest on record when judged by lay opinion or by the practice of his predecessors in the north. The expectations with which he had undertaken the cruise had been on the whole justified: the general course of the *Fram*'s drift had been north of west, curving upward toward the central Arctic on a line parallel to the supposed drift of the *Jeannette*'s relics. But the latter had been launched on their course advantageously at a point three hundred miles east of the ship's starting point. The arc described by the latter as she rounded the top of the globe was on an inner circumference and at lower latitudes, and further from the pole. The results obtained in oceanography and marine biology had already determined the success of the voyage as a scientific venture, but to extract all possible advantage from his effort Nansen proposed that early in spring two men should quit the ship and make a dash toward the pole by dog sledge—confident that they could effect a retreat, not to the ship, which would have shifted in a direction that could not accurately be predicted, but to land itself. After making the highest possible northing the two men were to return over the ice to Franz Josef Land, and travel by kayak to Spitsbergen through seas where they counted on enough ice to provide resting places and shelter for their frail craft in stormy weather. Once at

Spitsbergen they could count on being picked up by a Norwegian sealer.

Nansen found his crew a more sympathetic group than the sages of the Royal Geographical Society. His plan was endorsed with no scepticism and with no complaints, save from one or two applicants denied a share in the adventure. The two members of the traveling party were fixed on with little difficulty. Of the two leaders, Nansen and Sverdrup, one must go and the other must stay with the ship. Nansen very properly elected to travel; he was the younger man of the two, and had the medical knowledge almost certain to be needed on a long journey of incalculable privation and hardship. The Reserve Army Lieutenant Johansen was an easy choice as his companion on the journey over the ice.

Later A. W. Greely was to censure Nansen for having deserted his crew—a perverse and foolish comment—for within the meaning of the convention which binds captain to crew it was Sverdrup, the qualified master mariner, and not Nansen, the landsman scientist, who was to be regarded as the captain of the *Fram*. It would have been ridiculous affectation for Nansen to suppose that by staying with the ship he could contribute essentially to the discipline and safety of her crew on the way home.

Nansen anticipated Peary in the conviction that journeys such as theirs should not be deferred until spring for milder temperatures but begin as soon as the returning sun afforded light and guidance. He and Johansen quitted the *Fram* on 28 February 1895, fourteen years to the day before the American set out from Cape Columbia on Ellesmere Island on his final thrust for the top of the globe. But Nansen was taking no avoidable risks on a journey which, at best, was dangerous and uncertain in outcome. Twice he turned about and returned to the ship to make needed adjustments in the quantity and arrangement of his load. The date of his final departure was March 14. The *Fram* then lay in latitude 84° 02′, longitude 110° 50′, fifty miles nearer the pole than Peary's jumping-off point at Cape Columbia. Cape Fligely lay on the way back three hundred miles distant, west by south. The travelers secretly cherished the hope that by registering a steady ten miles a day they might actually reach the pole and set their return course straight for Spitsbergen.

Their route lay through a wilderness of hummocks, weathered into rounded lumps and buried deep in snow. Despite a hopeful start and good progress for a few days the travelers soon gave up their extravagant hopes. The surface did not improve as they drew farther from land: flat stretches were neither frequent nor of long continuance. Toward the end of March they came upon a recently frozen pool and wondered how such could occur in winter within five degrees of the pole. They then came into an area where *all* the ice was of recent formation and thrown up by pressure into jagged peaks and ridges over which the laden sledges could barely be hoisted. The dogs began to tire and Nansen describes the necessity of cruelly beating the loyal creatures to keep them on the move. Sights taken on March 29 brought the unpleasant realization that they were only at latitude 85° 30′ when accurate reckoning placed them above 86°. A few more days forced the reluctant admission that they were being carried back on their tracks by the southward drift which had baffled Parry above Spitsbergen seventy years before. April 1 found them in an "endless moraine of ice blocks, young and old." Their outward journey ended at (by their computation) 86° 13′ north, and 98° 47′ east, 132 miles from the ship and 226 from the Pole. Cape Fligely, where they hoped to make their landfall, now lay 450 miles to the southwest. The disappointed travelers would have been even more mortified at this meagre success had they foreseen that in the coming November the trend of the polar current as it flowed round the north end of Franz Josef Land was to push their shipmates on the *Fram* to within a few miles of the northing which had cost them so much effort to attain. A longer period of exertion lay ahead, for to go back to the ship was out of the question. She must have shifted and a mile or two of rugged pack would be enough to hide her from their view. Their only course was to make straight for Franz Josef Land, recruit supplies with the game which they knew to be plentiful there, and travel in kayak by sea lane and ice floe to Spitsbergen. They hoped to accomplish this in the present season.

For some weeks rapid progress seemed to justify this expectation. Their southward course, slanted away to the west, appeared to run parallel to the lines of pressure ridge which had retarded their advance to the north. Easier traveling did not lure them to any slackening of effort. With the certainty that spring was at hand they drove their dogs to the utmost, making journeys of up to twenty miles a day. As the loads on the sledges diminished they killed the weaker of the dogs and fed them to their mates who soon grew accustomed to this cannibal diet. They butchered the poor creatures with a knife, a horrible task, says Nansen, but necessary, for he and Johansen found their united strength unequal to strangling a dog, and they dared not avoidably diminish their stock of ammunition. One dog, Nansen's favorite, ran away, but lurked near the caravan and finally rejoined it. "He was evidently ashamed of himself, and came and stood quite still, looking up at me imploringly when I took him and harnessed him. I had meant to whip the dog, but his eyes disarmed me."

Fine cold weather through April gave the traveling conditions desired by rapidly freezing the leads which did occur. On April 29, when they had completed two-thirds of their journey and reckoned that Petermann Land was not more than a hundred miles away, they were halted by finding across their track a broad open pool. Troublesome as this obstacle was, and the omen of more trouble to come, the travelers welcomed the sight of "open water and glittering waves," as a rest for their eyes, long dazed by frost. They followed the margin of the lake in the direction where it appeared to narrow and soon heard a grinding roar as the edges of the rift came together, churning up a heap of rubble in the clash. Over this men and dogs scrambled, dodging the ice blocks which were tumbling down from the mounting ridge. They now began to come upon leads filled with floating pieces so small as to cant under a man's weight or covered with a icy sludge which might rest on a solid bottom or might be afloat, threatening the rash traveler with a plunge into the icy water beneath. The kayaks could not yet be used with advantage, so they were obliged to stray back and forth in search of a convenient crossing, or to camp on the margin of a lead until it froze to a thickness that promised security.

They were journeying in ignorance of their exact longitude, for the watches of both men had run down and been set by guesswork. Fortunately Franz Josef Land was no inconsiderable target, and Nansen aimed at its eastern side to ensure that if he arrived at no land in latitude 82°, he would be safe in steering west to find it. As they advanced strong east winds set the ice in motion, disturbed their reckoning and aroused the fear that they might be swept past Franz Josef Land and so miss the game of which they were urgently in need.

On May 16 Nansen with a positive latitude of 83° 36' and a longitude estimated at 59° 55' east, was puzzled at the nonappearance of land. He was not aware that he was actually six degrees out in his longitude and that the location of the supposed Petermann Land of Payer lay not in his path but seventy miles to the west.

Difficulties mounted as they drew nearer to land. With warmer weather the snow which filled the hollows and recesses of the ice had grown so soft that the men sank to their hips while the starved and overburdened dogs floundered in utter helplessness. Dark "water sky" was now visible to the south; wild fowl appeared; narwhals and seal were seen in the ice clefts, but there was still no sight of land on the last day of May.

On that date the appearance of an extensive body of water ahead necessitated a halt of some days to put the damaged kayaks in order for the seaborne traverses that they so ardently desired.

While they were thus engaged westerly winds brought ice toward them in thickening quantities and packed it in an unending field quenching the hope of easy transport by water. So the weary travelers replaced the kayaks on sledge and set forth again hauling with the few remaining dogs. For weeks they toiled through a morass of tumbled ice dotted with pools and still covered with snow which clung to the sledge runners like glue. June 21 brought relief from their most urgent fear. They had just made a water crossing and were hauling up the kayaks when a splash gave notice of a seal nearby. They secured the prize with rifle and harpoon, hauled it up and "launched out into excesses" on the blood-stained ice.

Being now relieved of the immediate threat of starvation the two men resolved to give up the frightful exertion which brought only negligible gain and to camp where they were until the snow melted down to firm ice, or the ice itself dispersed to give uninterrupted passage by boat. The capture of a polar bear and two cubs enabled them and the two surviving dogs to prolong their stay for a month. The ice remained firm strengthening their belief that it was backed up against some not very distant land.

NANSEN AND JACKSON IN FRANZ JOSEF LAND

However wise Nansen's decision to halt, it must have required plenty of resolution to adhere to it for several weeks, when every day he saw the sun slanting down closer to the northern horizon, serving notice that the peak of the year was past and another winter approaching. He and Johansen undoubtedly would have broken camp sooner had they known that all the time the land which they yearned for was within their range of vision. Away to the southwest, above the sodden grey of aging ice and snow, stretched a wreath of white, which Nansen more than once examined through his telescope. He could make out nothing but altering cloud shapes and stayed resignedly where he was, not suspecting that these apparent vapors, distorted by refraction, were snowy mountain tops of islets on the northeast of Franz Josef Land.

At last, on July 22, they decided that the snow was sufficiently dissolved for renewal of their journey. With a man and a dog harnessed to each sledge they made slow going, though the load was much reduced. Arriving at a lane of water Johansen paddled over in his kayak, with his dog stretched on the foredeck. Nansen, thinking his boat too cranky for so unstable a cargo, towed it over, stepping from one swaying elusive ice piece to the next.

On the morning of the twenty-third Johansen noticed a curious black stripe on the horizon ahead. Repeated disappointment had bred scepticism in him—he did not even mention it to his companion. Sometime later Nansen observed it and studying it through his glass he was soon assured that it was not a streak of clay on the ice close ahead, but a ridge of rock protruding from a distant snow field,—a barren spectacle, but to the Norwegians, who had almost despaired of seeing land again, an object of rejoicing.

On the twenty-fourth the travelers stepped out almost gaily, assured of reaching shore by the following nightfall. More land, crowned with a massive ice sheet, appeared to the west. It was Crown Prince Rudolf's Land, where Payer's journey to the north had ended. But though plainly in sight it was not easy to attain. Thirteen days were required to cover the distance which the travelers had reckoned on doing in one. On the twenty-fifth they worked hard with no perceptible diminution in the distance from the nearest shore. Then a wind arose from the south-southwest, opening lanes and driving them back several hard won miles. To add to their distress Nansen was now almost crippled with lumbago. By the thirty-first he was somewhat better, but they had been blown so far to the east that Rudolf's Land was out of sight below the horizon, they were lost in a maze of rotting ice pieces and water lanes so packed with brash as to be impassable on foot or by kayak. Long, and often fruitless, detours added to their exhaustion. "Inconceivable toil" notes Nansen for August 3; "We could never get on with it were it not for the fact that we *must*." With food in short supply they were obliged to shoot gulls to feed to the still indispensable dogs, expending precious ammunition for an inadequate return.

On August 5 they arrived at the margin of a navigable lead and busied themselves removing loose ice and pushing back the brash to launch the boats. Nansen had just eased his laden kayak into the water when from behind he heard a scuffle and a cry from his companion. He turned just in time to see a polar bear spring at Johansen and dash him to the ice. Nansen's rifle was in the floating kayak: he knelt and stretched down the icy brink to retrieve it. He heard a shout: "You must look sharp if you want to be in time." Never can a man have escaped violent death by a margin so narrow. Normally the beast would have broken Johansen's neck in a moment; but, startled at the strange creature that lay beneath him, worried by the dogs, and distracted at Nansen's movements, the bear faltered and had turned on the dogs when Nansen with a single shot laid him dead on the ice. Johansen had been stooping to lay hold of a sledge rope when he caught sight of a moving creature whom he took at first to be a dog; he just had time to recoil and lessen the shock of a blow that might have killed him. Nansen could never have pulled through without him, and, but for the savage creature's momentary hesitation, this man, for many years the first citizen of Europe, would have been merely one of the many gallant spirits, who bartering security for adventure, have vanished forever in "the dark places of the earth."

This lucky escape proved to be the dawn of better things for the two adventurers. Freed from the fear of hunger and strengthened by the meat they devoured raw on the spot, they pushed on by sledge and ferry and soon made out a dark stretch of open water ahead which seemed to extend to the shore of the island they had tried so hard to approach. On the seventh they reached the margin of an ocean strait, dotted only with floating ice and bounded by a towering glacier cliff on its opposite side.

It was out of the question to carry the dogs on an extended voyage by kayak, but not having the heart to butcher their favorites as they had their teammates, they agreed to sacrifice two rounds of ammunition, Johansen shooting Nansen's dog and Nansen Johansen's. They then lashed the kayaks together in order to carry the sledges laid crosswise fore and aft on the broadened deck.

As they headed west with sail and paddle they were perplexed, not knowing where they were. Nansen had supposed that they were approaching Wilczek Land. But at that latitude (which he *could* determine with precision) where Payer had shown continuous land, he saw, on the left, a group of islets (named by him Hvidtenland) and on the right, where Cape Buda Pesth and Dove Glacier should have appeared, bare and open sea. Deprived of these landmarks he naturally failed to identify high basalt cliffs raised to the northwest as the southern side of Prince Rudolf Island. Finding so little correspondence between what they observed and any part of Payer's map, the Norwegians hoped, though with little conviction, that they had underestimated their westing and were among the chain of islands which the British Leigh Smith had discovered extending west from the *Tegethoff*'s winter station towards Spitsbergen.

They coasted past Hvidtenland between floating pack and shore ice (Nansen named two islets of this group Eva and Liv after his wife and infant daughter), making the occasional portage where the pack pressed against the land. Steering by compass in heavy mist they crossed a broad stretch of ocean, the south end of Payer's supposed Rawlinson Sound, and made a landing on Houen Island off the north shore of Karl Alexander Land to enjoy the first dry bed that had been theirs for many months. They were now at ease in body and mind—they had found polar bear and walrus plentiful enough to assure food for the coming winter.

On August 17 they rounded Cape Hugh Mill and coasted to the southwest down the British Channel. A week later when approaching Salisbury Island they encountered storm, pack, and young ice, a deadly menace to canvas kayaks. The two travelers made mutual confession of what each had already admitted to himself—that they could not make Spitsbergen that season. They must establish themselves for the winter before cold and darkness put an end to hunting. They made their camp at the base of a rock strewn cliff on the south side of the island, later named Frederick Jackson in honor of the man who rescued them.

Though both men were desirous of exchanging their weather-worn and ragged tent for a stable and windproof hut, they first gave their attention to hunting bear and walrus. The latter were shot, secured by harpoon and towed to the shore or to fast ice. As the carcasses were too weighty to be landed they were skinned and cut up half-submerged, a disgusting task which drenched the hunters in water and steeped their clothing in blood and blubber. When they had gathered a sufficiency of meat and blubber, they fashioned a crowbar from a sawed-off sledge runner, and lashed a broken ski pole to the shoulder bone of a walrus to form a spade. They found a suitable spot on the sloping talus at the foot of the cliff, dug out earth and broken rock to form a basement three feet deep, quarried rock from the cliff to build up the walls of the hut, and gathered moss to caulk up the chinks. The lucky find of a piece of driftwood provided them with a ridgepole over which they could drape walrus hides, and finding the latter frozen stiff, they dipped them in seawater until thawed and flexible. They cached what remained of provisions brought from the *Fram* as providing the most portable supplies for the spring voyage to Spitsbergen.

So living on the flesh of bear and walrus and using blubber for fuel they passed the winter of 1895–1896. Their time was spent in cooking, eating and sleeping. They had no books except their diaries and little scope for exercise: their clothes were too steeped in blubber to give insulation; and in any case the winds blew cruelly around the cliff foot and their outer wind garments were too torn to be serviceable. Nansen's discovery that twelve threads could be obtained from one piece of cord came too late to be of much service; by the time his wardrobe was repaired the peak of winter was past.

The Arctic winter, often a trying period to parties far larger and better furnished than theirs, was passed by the two Norwegians in perfect good humor. They sometimes slept for twenty of the twenty-four hours. Some diversion was found in efforts to remove the coating of blubber soot in which both men were encased. Water was ineffective, rubbing with moss and sand nearly so. The best detergent they found was warm bear's blood rubbed in and scrubbed off with moss. When this was in short supply, "The next best plan was to scrape our skin with a knife."

The killing of the first bear of this season on March 8 relieved them from a threatened shortage of meat and blubber; with the rapid return of daylight both the men and their savage neighbors went abroad freely, and food was again in ample supply.

Though Payer had found open water in early April as high as Prince Rudolf Land, the ice in the vicinity of the camp remained unbroken to a much later date. As Nansen and his companion had little inclination to resume sledging with no dogs to aid them, they decided to remain where they were until the open sea drew nearer. The discovery that the stores buried the previous autumn had mildewed and spoiled seems not to have caused them much uneasiness. Still at a loss to determine their whereabouts without attributing serious error to Payer, they fancied they might be at the extreme west of the island group, within easy reach of Spitsbergen. But by the middle of May both men were growing impatient, Nansen especially dreading his wife's distress should the *Fram* reach Norway while he was still lost in the wilderness. On May 19 they set out sledging over the ice with the prospect of early arrival at the open sea that was to take them home.

While passing the winter in the darkness and squalor of their lonely hut and dwelling fondly on the delights of home, the two Norwegians had not suspected that barely a hundred miles to the southwest lay a camp where nine men were enjoying in large measure the civilized comforts which they did not hope to find on their side of Norway. British enterprise on Franz Josef Land dated back for more than a decade. In 1880 the wealthy Scottish globetrotter, Leigh Smith, had come up in the steam yacht *Eira* and, making his landfall to the west of the *Tegethoff*'s winter station, had begun the dismemberment of Payer's compact Zichy Land by the discovery of Bell and Northbrook Islands, and, further to the west, Alexandra Land. These were the lands which so possessed the imagination of Nansen as he was groping his way, in ignorance of his true longitude, down from the northeast.

Emboldened by this success Leigh Smith came back in 1881 to extend his discoveries. "But this time," says the Russian chronicler Viese, "the Arctic did not meet the travelers with so comradely a reception." The *Eira* was nipped and sunk off Bell Island: her crew, which had embarked for a summer excursion, were cast ashore on a low windy cape, with their boats, a little salvaged timber and two months' supplies—to support them for a period of not less than ten. Their case was nearly identical with that of Greely on Ellesmere Island two years later; lucky circumstance spared the British the all but complete disaster which overtook the American party. Their shipwreck occurred early in the autumn; the men were well fed and vigorous, and game proved to be in ample supply. Backed by his surgeon Dr. William Neale, Smith kept the men busy and hopeful; he built a comfortable hut, hunted polar bear, walrus and loons with great success, and throughout the winter employed all hands on preparation of sails, clothing and preserved meat for the coming voyage by boat. In the spring, warned by Payer's experience, he deferred his departure until late June, and despite much hardship, finished in forty days the voyage to Novaya Zemlya which had detained the Austrians for three months. Off Matochkin Shar he and his party were taken aboard by a ship sent to their rescue.

Leigh Smith's excursions had established two points which the voyage of the *Tegethoff* had left in doubt; namely that the Franz Josef Land was accessible to ocean-going vessels and that it furnished enough game to make its explorers self-supporting. Ten years later a brisk English sportsman-scientist, Frederick Jackson, formulated plans for the scientific survey of the island group. He obtained the financial backing of Alfred Harmsworth (Lord Northcliffe), a newspaper magnate of ambiguous celebrity, but the generous patron of discovery in both Russian and American polar sectors. Though both Harmsworth and Jackson disclaimed the purpose of reaching the pole, Jackson did nourish the hope of finding beyond Cape Fligely the northward-stretching chain of islets that Peary was then looking for at the top of Greenland; but scientific research remained the principal purpose of the expedition, and Jackson's staff, including meteorologist, geologist, botanist and mineralogist, was perhaps the most highly specialized team to undertake Arctic studies up to that date. Two of its members, Dr. Reginald Koettlitz, and Albert Armitage, lieutenant of the Naval Reserve, were later to go "south with Scott" to inaugurate research in Antarctica.

In 1894 the ship *Windward* carried the party to its destination. Jackson set up his base near Cape Flora on Northbrook Island, a hundred miles west of the *Tegethoff*'s old station and not far from Leigh Smith's winter camp on Bell Island. He relied upon Russian ponies for transport, setting an example which was to have a sinister influence on Captain Scott. Thus equipped, and accompanied by Armitage and the Finnish seaman Blomkvist, he set out by sledge on 16 April 1895 and went up a great sound stretching to the north—the British Channel. A hundred miles up he was stopped by ice so shattered as to be impassable to ponies, and turned back, baffled in his attempt to make a high northing but having made considerable corrections to the Austrian survey. In July he manned the whaleboat with all his staff except Dr. Koettlitz and set out to make scientific survey and to add detail to Leigh Smith's map of the western islands. Eighty miles out they were caught in a gale off a steep glacier which offered no shelter to man or boat, and for forty-eight hours (July 29–31)—while Nansen and Johansen were toiling toward the northeast of the archipelago—their future rescuers were in a more dire emergency on its southwestern side, fighting wind and surge with sea anchor out, vexed by cross seas, pelted with snow and sleet, and bailing for dear life. At last a lull in the gale permitted the chilled and starving seafarers to get their boat under control, weather Cape Grant and find a safe haven in its lee. There they got out their port wine to toast their boat and its skipper Lieutenant Armitage, whose "nautical knowledge and experience was of considerable service to us," says the matter-of-fact Jackson. "All my fellows have behaved well, and if we had gone to the bottom would have done so as becomes men."

Several of the ponies died in the winter of 1895–1896, and it was with a composite team of horse and dog that Jackson with the same two sledge mates set out on March 18, hoping by an earlier start to find the ice of British Channel firm and safe. In this he was disappointed; he was stopped by water some thirty miles short of his previous northing and about the same distance from the hut where Nansen and Johansen were even then passing the winter. Jackson was luckier than he supposed. By going north he would merely have been duplicating observations already made by Nansen: turning east as he did, toward Austria Sound, he was able to prove Payer's Zichy Land a nest of islands, before fighting his way through foul weather and rotting ice back to his Cape Flora base. He was quietly employed in finishing his maps when Nansen and Johansen left their hut and began to descend the British Channel.

The Norwegians set out with their kayaks on sledge hoping soon to be afloat. But westerly winds began to fill their open sea with broken pack and by a disagreeable paradox they were impeded by ice in May where Jackson had been stopped by water in March. Moving with difficulty over rotten ice and sometimes halted by stormy weather they spent four weeks in traveling the distance which the Englishman in reverse direction had covered in eight days. They kept the islands of Zichy Land close on the left until they found the shore of Hooker Island falling away to the southeast, and then set a course for land visible to the southwest on the other side of the channel.

Nansen, who had still hoped that they were on the Spitsbergen side of the Franz Josef archipelago, now had an inkling of his true position. Observing how easily a sunlit fog bank reaching from headland to headland could be mistaken for continuous shoreline, he now understood what had seemed to him grotesque errors of Payer's map; and on rounding Cape Barents he noted features resembling Leigh Smith's description of the Northbrook

Island south shore. The Norwegians had coasted to within four miles of Jackson's Elmwood House when an encounter with a clumsy or hostile walrus compelled them to make camp on the shore ice and repair Nansen's kayak. They slept and rose to prepare breakfast before reembarking on their voyage. They did not know that while so engaged Lieutenant Armitage was watching them by telescope from the top of Elmwood House, or that Jackson was even then weaving his way toward them among the hummocks which lined the beach.

Nansen, strolling outside the tent, heard distinctly among the myriad bird voices the bark of a dog. He listened and heard it again. He called to Johansen who came out and listened sceptically, but the sound was not repeated. Nansen told his companion that he could believe what he pleased, but he himself was walking ahead to meet the strangers, and testified to his conviction by pouring into the soup all that remained of their Indian meal, not doubting that he would taste civilized food before the day was out. Leaving Johansen to pack their effects he set out along the shore ice. Any doubts which assailed him as he tramped along were soon dispelled by the sight of tracks, too large for a fox's, too small for a bear's—evidently those of a dog. He speculated on the character of the strangers; a Norwegian sealing crew or foreign scientists?—he had learned of Jackson's plans before the *Fram* sailed. Among the hummocks there echoed the voice of a man calling his dog. The Norwegian listened: the cry came again, in English. The stranger soon emerged from the hummocks and responded to Nansen's halloo; washed and shaved and neatly arrayed in a check suit, he presented a lively contrast to Nansen with his greasy ragged attire, his features blackened with soot, his hair and beard long and matted. Jackson came forward with undisturbed composure and shook hands with the words, "I am glad to see you"; on being informed that he was addressing Dr. Nansen he civilly amended his greeting to "I am *damned* glad to see you." He took Nansen to the hut, whose inmates heartily cheered him before setting out to bring in Johansen. The latter guessing what had transpired ran up on his sledge the flag of Norway which was saluted in its turn. He came into camp in state, walking beside the sledges while others hauled.

When the flurry of greeting had subsided the Norwegians noted a certain reserve and embarrassment in the bearing of their hosts. This was soon explained: meeting them in such disreputable plight the Englishmen had naturally concluded that the *Fram* had gone down in the ice and that Nansen and his companion were all that remained of those who manned her. They were promptly informed that the two had voluntarily undertaken their tramp through the polar wilderness, and that their ship was safe and confidently expected to reach home as soon as they. Jackson gave Nansen a packet of letters which he had brought from Norway two years before and faithfully carried with him on all his journeys on the slight chance of the meeting now happily realized. Scrubbed, shaved and clothed as Europeans the Norwegians fell in with the camp routine, Nansen hunting and geologizing with his hosts, when not occupied in adjusting his map to Jackson's; Johansen, who had no English, grumbling at the nonappearance of the expected *Windward*, and wishing that they had persevered in their voyage to Spitsbergen. Jackson, who had narrowly escaped drowning in the western sea in a craft much sturdier than theirs, confided to his journal the conviction that they would not have made it. On July 26 the *Windward* arrived, and the Norwegians took leave of their host, who, with some replacements on his staff, was staying over for a third season.

Before he embarked Nansen held with Jackson a dialogue which might have provided wholesome reading for certain vehemently competitive explorers of the next two decades. He must, Nansen explained, bring out his travel book at an early date while popular interest was at its peak; and the book must contain a map tracing his route and embodying his latest discoveries. The observations he had made from Hvidtenland to

Cape Mill were his own exclusively, but how could he deal with the chart of the British Channel from Cape Mill to Northbrook Island where their surveys overlapped and Jackson could claim priority of discovery? Should he include it or leave the region of the British Channel a blank? Jackson's hesitation shows that he would gladly have been the first to communicate what he had been the first to find; but after some deliberation he gave Nansen leave to print his map unexpurgated and added an invitation to assign his own names to the features they had jointly discovered. To the Norwegian's objection that this *was* improper, he answered that it was a personal matter which concerned no one except themselves. There the discussion ended, but Nansen retained his scruples and in preparing his map the only use he made of the Englishman's liberality was to assign to the island where he and Johansen had wintered the name of *Frederick Jackson*.

The *Windward* carried the Norwegian adventurers to Vardo where they went aboard the yacht of an English tycoon and sailed on to Hammerfest. There Nansen received a wire notifying him that the *Fram* had arrived at Tromsø the day before.

The latter part of the ship's drift had been no more eventful than the first. She was almost stationary during the summer of 1895; in the winter the movement accelerated and, as already noted, a northward slant in the current bore her up close to latitude 86° north. In the summer of 1896, when the ice broke up north of Spitsbergen, Sverdrup attempted to extricate the vessel and force her to the south, and after weeks of labor broke out into the open sea. He put into a Spitsbergen bay to enjoy the welcome of Norwegian sealers and set a course for Norway where his arrival almost coincided with that of Nansen and Johansen on the *Windward*.

Jackson, for his part, stayed over at Cape Flora for a third winter with the purpose of making by spring sledge journey the complete coastal survey of the western Alexandra Land which he had failed to accomplish by boat. The outcome was a tour of seven weeks (March 16–May 7) which merits a higher place than it now possesses in the history of Arctic endeavor. Only Jackson and Armitage traveled; for Blomkvist had gone home on the *Windward,* and they foresaw more difficulty than advantage from adding a novice to their team. They were troubled by extreme changes of temperature and by violent storms which kept them weather-bound and threatened to shatter the frozen sea. Their one surviving pony and sixteen dogs were badly rundown: the pony and most of the dogs died on the trail.

Their course took them up the British Channel until they veered west through Leigh Smith Sound. They were much hampered by crushed up ice and pitfalls loosely shrouded in snow. Camping one night off a glacier face, they were nearly buried in a descending avalanche; the sea ice bowed under the load and nearly gave way to plunge them into the water beneath. After traversing Leigh Smith Sound and ascertaining that Alexandra Land was probably divided by an extension of Cambridge Bay they took to low land and then to the glacier top to bypass open sea, and so reached Cape Mary Harmsworth, seen by boat the summer before, and determined that it was probably the western extremity of Franz Josef Land. For a time they were able to steer direct for their home base across Weyprecht Bay, but found water and broken ice, impassable to sledges, reaching to the foot of Cape Ludlow. As the glacier face was quite inaccessible they cast about until they found a snow slope easy enough for men and dogs to claw their way up dragging sledges after them. From the glacier top it was seen that east of Cape Ludlow lay a large body of open water necessitating another wearisome detour northward to find a safe passage across the ice of Cambridge Bay. After making the descent from the glacier top—as difficult as its ascent had been—the travelers found fair going on sea ice, though Jackson occasionally climbed a headland to survey his intended route and ascertain where the open water lay. After relieving their craving for tobacco with cigar ends found in Leigh Smith's abandoned hut they made their last leg from Bell to Northbrook Island and reached Elmwood House on May 7.

When the *Windward* returned to take the entire party home Jackson prevailed on her captain to cruise west fifty miles beyond Cape Mary Harmsworth. He was thus able to confirm his opinion that that cape was the end of the archipelago and no more land lay to the west. Together he and Nansen had given Franz Josef pretty much the form it has in the latest atlas.

ANDRÉE'S BALLOON FLIGHT TOWARD THE POLE

About the time that Frederick Jackson wound up his affairs at Cape Flora and embarked for England three other adventurers launched off over the waters he was quitting on an enterprise as ill-calculated and rash as his had been methodical and deliberate. Sverdrup had had notice of their purpose the year before: when the *Fram* on emerging from the ice spoke to the sealer *Sostrene* her crew were told, along with other items of news, that even then the Swedish aeronaut Andrée was on Dane Island (off the northwest angle of West Spitsbergen), intending to take off by balloon and fly over the pole to Siberia. It so happened that unfavorable weather obliged him to defer his flight to the following year.

At this time Salomon August Andrée was forty-three years of age. He held the appointment of Chief Engineer in the Swedish Patent Office, was a practical scientist of note and well acquainted in the learned circles of Stockholm. For some years he had been interested in aeronautics and had made experimental flights by balloon, one of which terminated in his rescue from a Baltic islet by a passing ship. Notwithstanding this demonstration of the balloon's unreliability, he conceived the idea of a transpolar flight, communicated it to the great Nordenskiold, and encouraged by him, published his purpose and solicited subscriptions to enable him to put it into execution.

It is an odd circumstance that the polar experts appear to have divided for or against Andrée's plan in much the same proportions as they had aligned themselves in favor of or contrary to Nansen's bold experiment in the *Fram*. Men of science though they were, they would seem to have been governed more by enthusiasm than by calculated judgment in a particular case. There was no comparison between the two schemes in practicability. Nansen was taking one measured step into the dark, Andrée, an uncontrolled leap. Nansen in his admittedly risky undertaking was simply taking over where De Long left off with what he rightly supposed to be a trustier instrument; Andrée's venture was without precedent and his means of transport imperfectly tested. Lengthy flights by balloon had been made *over land* up to 800 miles by aeronauts who were tied to no specific destination and whose safety was assured by the machine's capacity for a "soft" landing in the smallest of spaces. Andrée wished to steer a predetermined course over 2,000 miles of polar sea where a soft landing might do no more than defer his fate. Those who encouraged him may unconsciously have shared the view which Colonel Watson of the British Aeronautic Corps somewhat callously expressed: "It may be that Mr. Andrée will never come back, but in spite of the risk the attempt ought to be made." It must be remembered however, that the gallant Swede was not committing himself absolutely to the caprice of variable and unpredictable winds. It was the opinion of his chosen companion, the meteorologist Ekholm, that a strong south wind could be trusted to take the aeronauts up to the pole where, after some eddying around, they might hope for a wind to take them down to the Siberian shore. Furthermore a method had been found for applying some control to the course of the wind-driven balloon. When traveling free, at the speed of the wind, it could not be steered, but by means of cables, "draglines" or "guide-ropes" dragging in the water or over the ice from a height of 400 feet it could be slowed down and, by the use of sails, its course deflected as much as twenty-seven degrees from the direction of the wind. Also, as the central Arctic was then quite unknown, Andrée could hope, in event of mishap, to find islands providing wildlife as stepping stones to safety. These considerations might not have weighed much with one in whom the faculty of prudence was dominant; they were enough for a bold and determined visionary. After all, Andrée's venture was hardly more desperate than the voyages of his Norse ancestors; he was truly the last of the Vikings, a reincarnation of Bjarni Herjolfsson and Leif Ericson, and one can only regret that he was not to have the luck of his transatlantic predecessors.

With the help of the wealthy inventor, Alfred Nobel, Andrée was able to raise the funds needed for equipment and general expenses. The former consisted of the balloon itself, canvas boat, three sledges, tent, arms and ammunition, scientific instruments and provisions for four months. The aeronauts hoped to reach Bering Strait in six days; but, apart from the chance of mishap on the way, their touchdown in Siberia might be far from any settlement that could afford them relief. As his companions Andrée chose Ekholm and Nils Strindberg. In the summer all sailed for the departure point in Spitsbergen. They waited for some weeks at Dane Island for the strong southerly wind needed to launch them on their course; when it failed to materialise, the attempt was put off until the next season. In the meantime Ekholm developed scruples and doubts, especially with regard to Andrée's estimate that the balloon was tight enough to keep its buoyancy for thirty days. He therefore withdrew from the expedition and very naturally advised Strindberg to do the same. But the poor young man, though engaged to be married and deeply attached to his fiancée, Miss Anna Charlier, felt that he had given his word to Andrée and could not in honor retract it. A civil engineer, Knut Fraenkel, volunteered to replace Ekholm. In 1897 the three aeronauts again sailed to Spitsbergen and took up their station with ship and ground crew awaiting the conditions for a safe ascent. On July 11 a southerly wind was blowing, strong enough, so the eager Strindberg and Fraenkel thought, for their purpose. Andrée, with some misgivings, signified agreement and gave the order to take off. As the balloon mounted up into the wind the guide ropes parted—the limited control they might have afforded was lost. Climbing higher as she crossed the water the *Eagle*, as the balloon was called, cleared a projecting spur of mountain and sank from sight in the void beyond. A carrier pigeon came back; years later buoys released by Andrée in the first hours of his flight floated to land. For more than a generation no other word of those gallant spirits came back from the icy tracts into which they had so rashly ventured.

They were not much remembered in the busy and eventful world they had left behind. Peary ended his twenty years' siege of the North Pole with an approximate success which the world gladly hailed as a complete one; Amundsen, Ellsworth and Byrd used the aeroplane to cross with ease the barriers which had thwarted sailors since the time of Henry Hudson; Amundsen and Nobile crossed the pole in the powered airship *Norge*; Nobile in the *Italia* surveyed great tracts of the frozen Arctic before crashing on the very ice over which the *Eagle* had disappeared. *His* catastrophe caused a tremendous stir in the sensation-hungry world of 1928; probably few of the hundreds who hastened to his rescue spared a thought for the first airborne discoverers of them all who lay obscure and all but forgotten almost on their track.

In the summer of 1930 the Norwegian sealer *Bratvaag* left Tromsø for a hunt in northern waters. In addition to her crew she embarked a number of scientists headed by Dr. Gunnar Horn. The cruise took her up the east side of the Spitsbergen group and on August 4 she approached White Island, an outcropping eleven miles in length lying ten miles off the Northeast Land. It is almost completely icecovered and rises to a cupola 660 feet in height, "a dazzling white shield floating on the waves." On the fifth the *Bratvaag* anchored half a mile from the island's southwest shore and sent her boats away on a walrus hunt. Late in the day Captain Eliassen came back to the ship to inform Dr.

Horn: "We have found Andrée." A boat's crew had pulled into a beach on the northwest side of a rocky hill to skin a walrus. On the beach they found first an aluminum lid and then a canvas boat canted over and filled with ice. Its contents, instruments, firearms and a parcel of books had been scattered by bears. The appearance of detached human bones led them to a skeleton reclining in a natural posture against a slightly sloping wall of ice. Its lower limbs were still encased in Lapp boots. The upper part of the skeleton had been pulled apart by bears and the head was missing. Marks on the clothing identified the bones as those of Andrée. Further search disclosed a grave thirty yards from Andrée's last resting place. In it lay the partially dismembered skeleton of Strindberg.

A rising wind, which made the open roadstead unsafe, compelled the men of the *Bratvaag* to collect what relics they could and make for the open sea with the search incomplete. Fearing to injure the contents of the boat—and possibly the still missing body of Fraenkel—by hacking out the ice, they dug it out of the snow and hoisted it and all it might contain bodily on board before weighing anchor.

On the report of this find journalists of various nations chartered the sealer *Isbjorn* for a voyage to White Island. With the summer thaw far advanced they found Fraenkel's body lying on its side two yards from Andrée's last resting place. Strindberg evidently had been the first to die and had been buried under a pile of stones. Andrée and Fraenkel had perished of cold, asleep perhaps in the shelter of their now-vanished tent. Their records, journals, logbook, and a shorthand diary written by Strindberg for his sweetheart, were for the most part still decipherable and furnished the story of all but the last few days of their tragic adventure.

The first few hours after the takeoff from Dane Island had been prosperous. The *Eagle* mounted to a considerable height and was carried north-northeast at some fifteen miles an hour by the wind. As the day declined she began to lose buoyancy and despite the unloading of ballast sank into the fog where accumulation of moisture necessitated the sacrifice of more ballast. The wind died away and she remained stationary, poised low over the silent, unresponsive ice fields. Then the wind rose again from the east and drove her to the west slowly—and this was fortunate, for by now the *Eagle* was riding so low that from time to time the car would bump on the top of a hummock passing beneath her. Again the wind failed and after a period came back from the west carrying the balloon and its helpless crew eastnortheast. Buoyancy was low and bumps frequent. In the small hours of July 14 Andrée began to consider ending his now hopeless endeavor before they were carried too far from home. At 7 A.M. he did so: he opened his valves and brought the *Eagle* down to the ice in 82° 56′ north latitude, thirteen miles north and some seventy miles to the west of Parry's turnagain seventy years before. The site of Nobile's destined crash lay a hundred miles directly south. Spitsbergen was 192 miles southeast.

These distances were not in themselves formidable to a healthy party as well provided as they. But the warmth of midsummer, which had made travel by balloon feasible, had produced soft ice and wide leads—deadly obstacles to men who must now journey on foot. Furthermore they were too few in number for the easy handling of the indispensable minimum of traveling equipment.

Andrée and his companions knew that a small depot lay at the Seven Islands just to the north of Spitsbergen and a large one at Cape Flora on Franz Josef Land. They elected to make for the latter. Perhaps with the irrational optimism to which men in their situation are prone, they hoped that Jackson had deferred his departure for another season and would be there to receive them. For some reason, doubtless the difficulty acute with so small a party of deciding what they could afford to carry and then packing it on sledge, they did not get in motion until July

22. Progress was slow as they were frequently obliged to bridge
leads, or, in the event of wider channels, to ferry their three
sledge cargoes over by canvas boat, a process involving two un-
loadings and two restowings for every lead. On one day in a
region of shattered ice 1,000 yards of distance was the meagre
return for a whole day of labor. Of necessity they shot polar bear
and seal to eke out their food supply: the fresh meat caused a
weakening dysentery which further retarded them. By the end
of July it was evident that a westerly drift was carrying them off
course and making Cape Flora unattainable. On August 4 they
altered course for Seven Islands. Wind, weather and the motion
of the ice were still against them: in the period September 6–9
they were carried back on their tracks for a distance of eighteen
miles, a loss of many days at the rate they were traveling. It soon
became evident that in the proximity of Northeast Land the permanent ice flow was deflected southward and was bearing them
into the open sea between it and the Franz Josef group. The tiny
White Island, just visible to the west, was a most uninviting
refuge. So the three travelers elected to winter on the drifting
floe, as the Americans of the *Polaris* had done in Baffin Bay; they
built themselves a comfortable ice house and soon killed enough
seal to support life until February. In early spring they could
hope to reach land over a solidly frozen sea. Here the connected
narrative of Andrée ends. On October 3 disaster struck. Some
commotion in the ice shattered the floe and wrecked the hut
which sheltered them. "Exciting situation," is Strindberg's
laconic comment for October 3–4. From Andrée's notebook
we learn that they landed on White Island on the fifth. An entry
of the seventh mentions the projected building of a hut. Thereafter his diary is illegible. The fragment of an entry by Strindberg, dated October 17, ends the pitiful tale. Strindberg, who
alone received burial, was evidently the first to die: Andrée and
Ekholm cannot long have outlived him as they had barely begun the construction of a hut. From the posture of their bodies

it appears likely that, worn out with dysentery and numbed with cold, they died in their sleep. Thirty-three years later the chance visit of the *Bratvaag* coinciding with a summer of exceptional warmth brought to light a tale that would otherwise never have been told.

THE ITALIANS: A THRUST FOR THE POLE

Parry's northern enterprise of 1827 remained for many years a story without a sequel. Then, toward the end of the century, the adventurers of the western world, with the older problems of the Northwest Passage and the source of the Nile solved, turned their attention to the North Pole, the attainment of which became a sporting event of popular interest and a source of competition among the explorers of rival nations. The American C. F. Hall provided the impetus for these expeditions with his 1871–1873 *Polaris* voyage, which, but for its leader's untimely death, might have further extended its considerable achievement. The British Nares' expedition, following in Hall's track with the professed purpose of walking to the pole, quickly bogged down (if that phrase is permissible) among the mammoth hummocks of the Lincoln Sea. De Long's bold reconnaissance in another sector inspired the *Fram* voyage which still left the central polar cap unattained. So while the dogged and single-minded Peary pursued his elusive destiny on the north Greenland and Ellesmere shores, others gave attention to Franz Josef Land and Cape Fligely, lying almost as close to the pole as Peary's jumping-off point, and of far easier access from home ports of Europe. But no Eskimos, the indispensable element in the American's success, were to be found there. The sane and practical Jackson preferred solid scientific work among the islands to pursuit of the polar vision. Nonetheless an Italian expedition was launched two years later to make Prince Rudolf Island (or a more northerly land) the base for a journey to the pole.

The Italian venture was unique in that it was the only polar journey of discovery to be headed by a member of one of the royal families of Europe. Its commander was Luigi Amedeo of Savoy, Duke of the Abruzzi and an officer in the Italian Navy. He was twenty-six years of age. His second-in-command and head of the scientific staff was Captain Umberto Cagni. Other Italian officers were Lieutenant Francesco Querini, and the surgeon Dr. Achille Cavalli Molinelli. The Norwegian sealer *Jason* was purchased, and renamed the *Stella Polare*. Her former skipper, Captain Julius Evensen, was retained to command the ship.

The *Stella Polare* sailed from Christiana (Oslo) 12 June, 1899, called at Archangel for dogs, and made her Franz Josef landfall at Northbrook Island on July 20. Since he was cut off from British Channel by ice in Nightingale Strait, Abruzzi steered west along the shore where Jackson's boat crew had almost come to grief, but thwarted in his purpose of rounding Cape Mary Harmsworth to the north, he came back and with much trouble pushed up British Channel past Prince Rudolf Island into the waters above Cape Fligely. No Cape Sherard Osborn or Petermann Land was to be seen affording hope of a high northern anchorage: evidently Jackson had missed nothing by his apparent lack of enterprise. Rather than go south and so lengthen the journey toward the pole Abruzzi berthed his ship on the west shore of Prince Rudolf Island in Teplitz Bay, an open bight and unsheltered from the west and south. On August 27 a gale from the southwest tore the *Stella Polare* from her anchorage, drove her broadside onto the land and piled her up on the shore ice. As the pressure relaxed she canted over seaward and passed the winter ignominiously guyed to the shore by steel cables to hold her from tumbling over on her broadside. Perhaps the mishap was inevitable: it could not have occurred in a more benign form. The damage to the ship's hull was capable of repair; her propeller was uninjured; and resting not on the sea bottom but in a dry dock formed by the ice, she was insured against the complete disaster which was to overtake Fiala's *America* in the same bay.

Camp was set up on the beach, storehouses and shelter for the dogs provided, while the crew passed the winter insulated from the cold in double tents. Even in the depth of winter men and dogs were trained diligently for the spring journey. On December 23, Abruzzi and Cagni, out on one of these exercises, were caught in a blizzard. Owing to a misplaced reliance on the instinct of the dogs to guide them back to camp, they lost their way, tumbled down the twenty-foot face of a glacier onto the beach, and reached camp exhausted and frostbitten. Abruzzi's left hand was so badly frozen as to be powerless and very susceptible to cold; in consequence he was obliged to relinquish to Cagni the command of the expedition to the pole.

This was to take place late in February. When once away from land and launched on its journey the polar party was to be made of three detachments of three men each, two of them supporting parties commanded by Lieutenant Querini and Dr. Molinelli (usually referred to in the record by his middle name, Cavalli) which were to supply the third team, made up of Cagni and two others, for the first stages of the outward journey and start it off on the final leg with its own food store intact. It had been intended to drive straight out to sea from Teplitz Bay, in order to get away quickly from the coast where ice disruptions were most prone to occur; unluckily just before February 19, the day appointed for departure, a disruption did form a lake at the foot of Cape Saulen; and the sledge crews, like Payer himself, were obliged to climb up and over the glacier top and make their descent to the sea west of Cape Fligely.

Two days' travel over the sea ice convinced Cagni that there were a number of defects in the organization of his column, the sledge loads in particular being too heavy for conditions encountered. Rather than persevere and court failure when it was too late to renew the attempt, he put about and returned at once to Camp Abruzzi. In a phrase quaintly reminiscent of a campaign of Louis XIV, we read that His youthful Royal Highness graciously approved of the veteran captain's promptness and decision. March 10 was set as the new date of departure, loads were adjusted and also reduced, as the period of the journey was now of necessity reduced from ninety to seventy-two days. The main party now had forty days for the 500-mile journey to the pole as orders were to turn back not later than April 20. On the appointed date the little caravan launched out on Teplitz Bay. Their commander saw it round the now securely frozen ice at the base of Cape Saulen and crawl out of sight.

From the outset Cagni set a course a little to the east of north to counteract the westerly movement of the pack. He appointed two sturdy seamen, Guiseppe Petigax and Alessio Feroillet as "guides." It was their duty at the beginning of the daily march, while others were breaking camp and restowing sledges, to go ahead with one team, test newly frozen leads for safety, find and, if necessary, hack a way through the ragged pressure ridges that barred their course. Sometimes they came upon a northward trending lead, frozen firmly enough to permit rapid travel over a surface that was both smooth and secure. Such windfalls were more than offset by winds and bitter cold. More than once Cagni shortened the period of daily journey, that the men might rest and treat their frostbites in the shelter of the tent. On March 23 when the first supporting party was released Cape Fligely lay only forty-five miles away, barely sunk beneath the horizon to the south.

It had been Cagni's intention to send back with the first party Dr. Cavalli, who had not been enlisted as a dog-driver and had only won a place on the team by his vigor and natural aptitude for the work. He chose instead to send back Lieutenant Querini, who was frostbitten and losing strength, along with the Seaman Ollier, also in poor condition, and the engineer Stokken (selected because he was Norwegian, and it was wished to keep the advance party as nearly "all Italian" as considerations of fitness allowed). Two weeks of cold and toil had robbed the adventure of its bloom: the three were secretly envied by the others whose road still led away from the ship. After another week of travel Cagni directed the doctor to return, adding one of his men, Canepa, to his own team of Petigax and Fenoillet. He had given up hope of reaching the pole but was still resolved to break the high northern record of Nansen.

They advanced as before with Petigax and Fenoillet going ahead to find or build a road, while Cagni and Canepa loaded and guided the convoy of sledges. They found that the moderating temperatures of April were not pure gain, for leads were slow to freeze over. On one occasion the captain and Canepa were cut off from the two guides by a sudden rift in the ice and lost much time in making a crossing and rejoining them. On April 6 they were caught within the arms of a great horseshoe rent, with no road open except the one behind them. Rather than wait for the lead to freeze into a thickness that promised security, they cast about until they found a wreath of sludge welded into the young ice and over that precarious bridge they crossed to the northern side of the rift. Encountering another lead which did not offer this resource, they quarried ice slabs from a hummock nearby, threw them into the water and with poles prodded them into the form of a bridge. A day or two later they witnessed a striking sight when a moving floe ground into the sheet on which they were encamped:

This morning a strong pressure piles up, at not more than a hundred yards from us, a regular wall of ice from thirty six to forty five feet high. . . . Enormous blocks of ice roll down in our direction with a loud crash; after having been thrown up by other blocks, lifted to the brow of the ridge, as if they were straws and rolled down in their turn. Their fall raises a cloud of ice-dust which envelops and whirls around the base of the pressure-ridge. The loud and continual crashing of the pressure is drowned by the booming of the immense cascade of blocks, which makes the ice on which we stand tremble.

However such convulsions were rare and easily avoided. What most tried the nerves of the travelers was the crossing of a broad, freshly frozen lake. Though marching smoothly over a surface for the moment secure, they were painfully aware that they were running the gauntlet, that a high wind, putting the older ice in motion, would shatter the thin and brittle surface like glass, leaving them to perish without remedy.

Storm-bound on April 18 at about 85° north latitude, two days before the prescribed date of return, the four men held council and agreed to reduce their ration and make a supreme effort to reach the 87° latitude. They were obliged to settle for something less. After exceeding, with great rejoicing, Nansen's record they pushed on to 86° 34′ north latitude and there on 25 April, 1900 hoisted the flag of Italy. The record for three centuries held by the seamen of the north, Great Britain, the United States and then Norway, was now in the possession of a Mediterranean power.

Cagni's homeward journey (which was to prove as close-run an affair as Dr. Cook's obscure and devious wanderings among the Queen Elizabeth Islands or Shackleton's crossing of the Georgia Sea) opened auspiciously. For ease in travel and to spare the guides he went back along his upward trail, though aware that it must have been displaced and slanted to the west by the movement of the pack. When it could no longer be traced the party steered south southeast and for two weeks made good distances over ice of comparative smoothness. This, they found, was not wholly advantageous: for at the day's end they pitched the tent in a spot often unsheltered from the winds which blew unchecked over the level surface. When storm-bound on May 2 Cagni, who had long suffered from a frozen and suppurating finger, took the opportunity of scraping the decaying flesh from the bone, and finding the latter dead, but causing intense pain up to the shoulder when touched, he whittled it off with a pair of scissors. He spent two hours, he says, in what a doctor would have performed in three minutes, while his comrades watched in fascination and horror. "Canepa at one time could stand it no longer and fled from the tent, in spite of the wind and snow."

Though hungry and thinking of their homes "and a hundred acts of gluttony," the men were at first in good spirits; they had become adroit in the bridging of leads, and an observation of latitude showed that owing to the drift they were fifteen miles further south than they had supposed. "We feel that we can live free from anxiety," noted Cagni on May 8. The next day brought sad disillusionment: they entered "a wilderness of channels, of great and little lakes, and ice still new and weak." More disheartening was the discovery by longitude calculation that the drift had carried them more than fifty miles west of their intended course. Two days later, in spite of inclining more to the east, they were even further off course to the west. They were approaching the latitude of Teplitz Bay, but like Andrée, of whose fate, of course, they were ignorant, were being sucked past the land into the waters between Spitsbergen and the Franz Josef Islands.

For a month they tried hard with dwindling food store and failing strength to counteract the southwesterly drift and make the shore of Prince Rudolf Island. But difficulties mounted with the advance of spring. At times their way lay amidst closely packed hummocks, through "deep soft snow, which exhausts the men and the dogs. By dint of unheard-of efforts the sledges advance with difficulty by stages of fifty or sixty feet. We sometimes sink to the waist at every step for a distance of six to ten feet; then the snow supports us for some feet further, and then we sink into it again for the whole length of the legs or up to the armpits; and we pull ourselves up on our arms or knees only to fall again." At other times they advanced as if on stepping-stones through a network of narrow rifts and small islands of ice. The two kayaks they carried had been punctured and their ribs cracked from rough handling on the road; it was thought best to defer repair work and reserve them for the lengthy voyage which might lie ahead. Once they followed the edge of an impassable lead to the northeast, hoping for it to narrow; when instead it bent to the west, away from land, they camped on its margin until the eddying pack closed in and gave them passage. Confronted by another wide channel they embarked themselves, their dogs and other belongings on a piece just large enough to float under their weight; and Cagni, who had been for some time sick and inactive, but, now recovered, was the freshest man of the four, hopped across the lead on floating fragments which yielded under his weight and secured a line on the other side by which his comrades hauled themselves and their floating dock over to firm ice. By such exertions, protracted through several weeks, they barely offset the remorseless westerly drift; all the time they were floating past the refuge they longed for, and on June 8 lay thirty miles south of the latitude of Teplitz Bay. The four travelers now resolved on a course that they had long debated: since they could not successfully oppose the drift they would march with it and make a landing on Prince George or Alexandra Island. There, living off game, they could tramp to Nightingale Sound and cross by kayak to Cape Flora where the homeward bound *Stella Polare* would be certain to call before giving them up as lost.

On the morning of the next day June 9, Petigax climbed a hummock to survey the ice and shouted to Cagni that he could see land to the southeast. Cagni quickly confirmed this and identified the land as the islets of Harley and Neale. He concluded that his chronometer had been wrong and that they were at least one degree (ten miles at that latitude) east of their supposed position. On the eleventh the islets were lost to view, veiled perhaps by a light haze; on the twelfth in clear light, land stood out boldly and the supposed Neale Island was recognized as Cape Hugh Mill, just north of Nansen's hut—they were even nearer to safety than they had believed. Close to shore the rate of drift was slower. They reached the shore ice of Harley Island and, skirting it to the northwest, took again to the sea and on the fourteenth landed on Ommaney Island where they took "an infantile pleasure in touching earth." From there they took off again over "nicely balanced blocks of ice which turn upside down when we have crossed them," and headed for Prince Rudolf Island now looming high to the northeast. The twenty-first found them on a small ice piece afloat on a large sea. The pack behind was still moving south with the current, but a strong breeze was blowing from the southwest. Though the kayaks were not serviceable, their sails were: these the sailors spread on masts improvised from oars, and after a voyage of two hours their unmanageable craft crashed into the shore ice near Cape Brorok. The four men with their surviving dogs climbed to the glacier top, passed behind Cape Auk and from the unexpected quarter of the southeast descended on their startled comrades at Teplitz Bay.

The joy felt by Cagni and his comrades at this lucky deliverance was dimmed by the news that one of their supporting parties had been less fortunate. Dr. Cavalli had brought his detachment in safely, but Querini, Ollier and Stokken were still missing, and after the lapse of three months must be presumed dead. Theirs had been the shorter journey; but fifty miles in wind and bitter cold might well have overcome the endurance of men already judged wanting in fitness. The collapse of one, the refusal of his comrades to desert him, would have sealed the fate of all three. Freeing the embedded ship from her bank of ice proved no easy matter; the crew spent most of the summer and all their guncotton in extracting her. They got her afloat on August 10 and after a fruitless call at Cape Flora gave up their comrades for lost and steered for home.

THE AMERICANS: FIALA'S *AMERICA* IN FRANZ JOSEF LAND

The voyage of the *Stella Polare* to Franz Josef Land was followed by two successive American attempts on the pole from the same quarter, unsuccessful in their main purpose, but still of interest to the polar student, and not without benefit to science. In 1901 the first of these expeditions, financed by William Ziegler and commanded by one Baldwin, arrived at the archipelago in the converted whaler *America*. It failed to get much above latitude 80° and wintered at Camp Ziegler on Alger Island on the south side of Markham Sound. In the spring of 1902 Baldwin carried 40,000 pounds of pemmican to Prince Rudolf Island and cached it at Cape Auk four miles south of Teplitz Bay. He then returned tamely to Norway and left his ship at Tromsø while he and his party took passage for the United States. Undeterred by this failure Mr. Ziegler put up the funds for a second expedition and gave the command to Anthony Fiala of Brooklyn, a photographer, who had served with Baldwin in that capacity. Second-in-command and Chief Scientist was William J. Peters of the U.S. Geological Survey, and Assistant Scientist was artist Russell W. Porter. Personnel totaled thirty-nine, almost double the complement of the *Stella Polare*.

In the spring of 1903 they sailed for Norway and took over the *America* at Tromsø. With routine difficulties, but no serious mishap they passed Cape Flora and reached Teplitz Bay, their intended base (named Abruzzi Camp), late in August. When the ship's captain, Coffin, protested that he could not on his own responsibility winter his vessel in so exposed an anchorage, Fiala handed him a written order requiring him to do so. He naturally wished to keep the expedition united and under his control, and the ultimate safety of all was guaranteed by the relief ship promised in 1904. The skipper's fears were soon justified: in late October, when most of the company were safely housed on land a gale shattered the bay ice and drove the *America* out to sea. The maintenance crew got steam up and brought her under control, but, lost in the darkness and driving snow, they barely recovered the harbor before the sea again hardened into ice. In December the ship was reduced to a wreck by a nip, and soon after, her hulk was swept out to sea in a blizzard and seen no more.

The polar party set out from Camp Abruzzi on 7 March 1904 with pony and dog teams. Owing to the thinness of the ice in Teplitz Bay it took the land route, clambered over the glacier and came down to the sea near Cape Fligely. Then it bogged down. Leaking oil tanks were difficult to repair in the bitter cold; one man had suffered a rupture, another a strained back; three others declared themselves no longer fit for the work. A blizzard which kept them storm-bound for three days gave the malcontents time to enlarge on and propagate their discontents. When the storm blew over they went back to the camp to reorganize.

The second attempt, launched on March 23, broke down in the ragged pack east of Cape Fligely. The ponies, unlike the men, proved too willing and by their rapid pace and contempt of obstacles upset the sledges to which they were attached. Other sledges broke down because their loads, through indifference or malice, had been unevenly distributed. Three miles from land they came upon a lead, the sort of obstacle which only keenness and universal goodwill could surmount. Fiala acknowledged defeat and canceled the polar journey for that season.

The expedition was saved from complete and ignominious

failure by the loyalty of its principal members and the resolution of the commander himself. He sought and obtained a number of volunteers who consented to remain at Camp Abruzzi and share in a journey north in the spring of the next year. The Chief and Assistant Scientists, Peters and Porter, were left to command the post and do what scientific work they could, while Fiala, in conformity with his duty as commander, conducted the rest of the party to Cape Flora, either to see them aboard the relief ship, or to make provision for their safety through the winter, if no relief ship arrived. He set out with his band in April over the frozen sea and covered the 150 miles to Cape Flora in sixteen days. His enterprising subordinates, leaving six men at Camp Abruzzi in charge of Mr. Stewart, the Commissary, and Chief Engineer Hartt, embarked on extended field trips. Porter came south through the heart of the archipelago, and Peters at a later date coasted down the east side of the British Channel by canoe. Both ended their journeys at Cape Flora reporting scientific achievements which were to save the credit of the expedition.

The nonappearance of the relief ship that summer caused Fiala both disappointment and anxiety. He saw himself deprived of the helpers whom he had hoped to recruit from her crew and was concerned for the well-being of those who must now pass the winter at Cape Flora. They were fully supplied by Jackson's old stores and a depot left by Abruzzi, but Fiala dreaded the quarrels almost bound to arise among men who were bored, frustrated and without duty or responsibility. Porter, the ever willing, declared his readiness to take charge of the camp until February when he was to sledge up to Camp Abruzzi to share in the polar journey.

Fiala himself left on September 27 to rejoin his comrades in the northern camp. As the sea was unfrozen he was obliged to advance by island hopping, a course avoided when possible (Payer had well described the archipelago as "Alpine" in profile); it was difficult to find a way up the steep face of the glaciers which covered them, and still harder to find a safe line of descent to the sea. His party's passage over De Bruyne Sound was long delayed by churning ice, impassable to sledge or canoe, and they reached Hooker Island with the sun vanished and nothing but diminishing twilight to guide them. Fiala mounted with his comrades to the glacier top, and, relying with too much confidence on the assurance of quartermaster Rilliet, who had traveled that route with Porter, that there was no danger from crevasses, began the descent toward the sea without the precaution of roping the men together. Scouting ahead in the dim light Fiala suddenly felt the snow yielding beneath his feet. The steward Spencer dashed forward to grasp his hand and the two plunged down together. After an interval of unconsciousness Fiala came to to find himself wedged between the faces of the crevasse, below him a black void in which his groping feet found no support, and above black walls of ice barely visible against the light from the hole through which he had fallen. From below groans assured him that Spencer was at least alive: the frozen slab which had given way beneath him had become wedged in its descent and arrested his fall. Their comrades hauled them out of the "horrible pit" by singular good fortune with no bones broken. Fiala tramped along with the dog teams as before; and Spencer, after riding on sledge for a brief interval, did the same.

After being five days storm-bound at Camp Ziegler, Fiala took to the trail again on November 5. He left two men at the camp and also his canoes, in order to march with the utmost speed in the brief noon twilights. Open water still occurred in those restless channels. At the base of the glacier on Wiener Neustadt Island they found open sea, a great steaming black void sending up columns of dark vapor in the cold air. "We rounded its fearsome edge, . . . in an almost lightless night, our sled runners only a few yards from the water, and gained the solid ice of Collinson Channel." In three days, November 10–13, they made the difficult and perilous crossing of the sea east of Jackson Island. At times they were passing over young ice in serious danger of drowning, for if the ice broke up in a gale its fragments would be too light to support them; at times they advanced "like so many shadows" through rough ice in darkness so intense that they groped with poles for obstacles in their path and, says Fiala, "actually walk into icebergs without seeing them." They reached the security of Hohenlohe Island—thirty years before the rest camp for Payer's dropouts—when the dreaded storm overtook them; when it blew over they went on by moonlight, and on November 20 were reunited with the four men who had spent the summer at Camp Abruzzi. Delayed by storm, mishap and treacherous ice the travelers had been afoot for two months and averaged less than three miles per day on the road.

During the spring journeys Fiala had appeared too trusting and naïve for large scale organization and command. But his perseverance through danger and hardship in order to keep faith with the lonely few at Camp Abruzzi showed courage of a high order and a capacity for winning the ungrudging support of those nearest him. Peters and his four other companions might well have insisted on wintering at Camp Ziegler instead of groping forward in darkness: they knew that the men at Abruzzi were well supplied and had nothing but loneliness to fear; they probably viewed the intended third polar attempt as no more than a gesture to save face for their commander and the millionaire who sponsored him. Yet they followed him without complaint and actually urged him on when he felt scruples at the risks to which he was exposing them.

Their fidelity was matched by that of Assistant Scientist Porter who had been left in charge at Cape Flora. On 18 February 1905 he turned over his command to Captain Coffin and the Greely veteran, Francis Long, and with one companion, the Newfoundlander, Duncan Butland, set out with dog sledge for Camp Abruzzi, to take his part in the polar journey. As they crossed De Bruyne Sound the first blush of returning daylight appeared to the south. It slowly faded and a rising blizzard drove the two men to the shelter of an igloo close under Hooker Island.

When the twilight of midday returned they still heard the gale raging outside through the translucent walls of the hut and became aware of a dark shadow slowly rising all around it. Snow swept from the summit of the island was spilling down the glacier face and burying them in a mounting heap of drift. Porter climbed out through a hole in the roof, took the dogs one by one from Butland and then hauled his comrade up to the surface of the drift. Marking the position of the vault with a ski planted upright in the snow and carrying firearms and sleeping bags they moved on some distance and built themselves another shelter in a more secure spot. When the storm subsided and they came back for their belongings not a trace of hut or marker could be seen. It was useless to search for it in the featureless expanse of snow; their sledge and most of their food was lost. A timely encounter with two polar bears saved them from starvation; they harnessed themselves with the dogs to bags holding the meat and other belongings and so trailed the unwieldy burden forty miles to Camp Ziegler, where Rilliet had been left in charge. Reequipped, the undefeated Porter went on and reached Camp Abruzzi on March 17 to learn from Chief Engineer Hartt that he had come too late and that the polar party was already on the road.

Porter, as perhaps he anticipated, missed nothing by his late arrival. Fiala had got his caravan of men and dogs over the glacier top and fifteen miles out to sea when at latitude 82° he came upon a broad, brash-choked lead in a temperature too mild to promise a speedy consolidation. He spoke of going on himself with one man and three dog teams, when Peters, hitherto discreet, intervened to end his leader's indecision. At that date, he said, there was little hope of breaking Cagni's record. By trying to do so, Fiala would risk not only himself but the dog teams, which would be badly needed in the coming winter if, for a second time, the relief ship failed to get through. Relieved of a measure of responsibility by this sensible advice (which probably coincided with his own inner conviction), Fiala gave the order to return. They all reassembled at Camp Abruzzi and in small detachments trekked south to Cape Flora and there boarded a vessel destined to worldwide fame, the relief ship *Terra Nova.*

The scientific gains of the expedition were extensive enough to form a volume published under the auspices of the *National Geographic Society.* In addition, in his *Fighting the Polar Ice* Fiala gave the world as attractive a book on northern travel as any author, American or British, has produced. Unlike some of his polar contemporaries he does not give himself an image of toughness and insensitivity to hardship, but portrays the horrors of dark and fog-obscured icy wastes with a nervous force which those ponderous egocentrics, Peary and Stefansson, would have despised. He is generous in his appreciation of Peters, Porter and others who stood by him; with those who let him down he deals more in sorrow than in anger—and names no names. His book is enriched with photographs which modern equipment and techniques have not improved on.

The name of Anthony Fiala does not appear in the *Encyclopaedia Americana.*

THE NORTH POLAR FLIGHTS
OF AMUNDSEN AND NOBILE

In 1909 Peary attained the North Pole. A few years later the wintering crews of Russian icebreakers made the last noteworthy discovery of new land in the Severnaya Zemlya (North Land) beyond Cape Chelyuskin. About the same time Stefansson reached the northernmost fringe of the American Arctic archipelago. These achievements, coinciding with the invention of a new mode of transport, marked the end of privately conducted pedestrian surveys in the polar regions. This book must end with a brief narration of the early experiments with the flying machine in the Arctic which were a prelude to the most modern age of systematically organised national polar work, by means of the icebreaker the aeroplane along with the use of the drifting ice island which the development of wireless and air transport had at last made feasible and safe. Ronald Amundsen, the most eminent survivor of the old school of discovery, was also the earliest pioneer of the new.

Like Shackleton, Amundsen had attained the summit of success too early in life and for years was vexed by the search for employment worthy of his reputation and in keeping with the restless spirit within him. He spent years in acquiring means for some new endeavor and expended it in a long and relatively fruitless voyage through the Northeast Passage. The development of aeronautics in World War I and the subsequent crossing of the Atlantic both by airplane and dirigible gave him the idea of surveying the Polar Cap by the same means. Late in 1924 he met the wealthy American engineer Lincoln Ellsworth, who promised to furnish both the necessary funds and his own co-operation in putting the scheme into effect.

Together the two men drafted the plans for a transpolar flight from Spitsbergen to Alaska by dirigible. The powered airship was still the safest mode of air transport and, for the distance contemplated, the only one. It had a wider cruising range, and in event of mechanical breakdown, which would force an airplane to make an immediate landing, the dirigible could hover until the needed repairs were made. This capacity for remaining stationary while airborne promised to be of great advantage in the study of particular features of land or ocean surface. Furthermore the airship had space for a staff of scientists in addition to its working crew. For Amundsen and Ellsworth planned no vulgar publicity "stunt," but continuous observation of the unexplored global surface from Spitsbergen to the geographical pole and on over Stefansson's "Pole of Inaccessibility" to America, coupled with the most precise magnetic and meteorological studies that could be made. While negotiations were proceeding for the purchase from the Italian government of the airship N-1, the joint adventurers undertook a flight of reconnaissance as far along their intended route as an airplane could be trusted to carry them. Ellsworth provided the funds for the purchase of two Dornier flying boats and their transportation to King's Bay, Spitsbergen. On 21 May 1925, he and Amundsen took off for the north in their respective planes, each with a crew of two. Six hundred miles out and within two degrees of the pole the motor of one plane failed, and both made an emergency landing on a tiny lake in the heart of the ice fields. The only resource was to pack the six men in one flying boat for the return flight; as the lake refroze they were obliged to spend three weeks on the scantiest of rations leveling an airstrip of five hundred meters on the ice. From this the overburdened flying boat made a dry takeoff and bore both crews in safety to Spitsbergen.

With this reconnaissance accomplished the partners traveled to Rome to negotiate the purchase of the N-1, renamed the *Norge,* and to engage the airship's designer, Colonel Umberto Nobile of the Italian Army, as pilot on the transpolar flight. They then departed for the United States, leaving Nobile to direct the refitting of the *Norge,* and making Messrs. Thommessen, Sverre and Bryn of the Aero Club of Norway their agents for working out the business details of the expedition.

This arrangement was to give to a controversy, petty enough, and conducted with childish intensity. Amundsen had his full share of grasping egocentricity, an occupational disease to which the explorers of the era were particularly prone. He was determined that the flight should be Norwegian and made under the Norwegian flag. Ellsworth, as chief paymaster, had to be admitted to the honors of the expedition; but "I did not intend, however, to share them with the Italians." Though no one except Ellsworth could rightly object to this exclusiveness, it was displeasing to the Italians who had offered Amundsen the *Norge* as a gift if he would fly her under their flag. Nobile urged the moderate request that he should be admitted as a third partner in the overall command. The agents, who must have been embarrassed by Amundsen's crude discourtesy in treating a distinguished engineer "solely as a hired employee," obtained through Ellsworth—a less fervid monopolist than his colleague —certain ambiguous concessions which bore in them the seeds of war. Nobile's request for five Italian mechanics whom he could direct in his own language was granted without cavil.

In the spring of 1926 Amundsen and Ellsworth sailed to King's Bay, Spitsbergen, and thither the *Norge* with Nobile in charge, followed them, arriving on May 7. It was during the ensuing period of preparation for the final takeoff that Commander Richard Byrd of the United States Navy and his copilot, Floyd Bennett, made their nonstop flight by Fokker monoplane from King's Bay to the pole and back, May 8-9. Amundsen could afford to view this exploit with indulgence; for Byrd was doing little more than duplicate his own flight of the previous season.

On May 11 the *Norge* took her departure from King's Bay. Her crew of sixteen was made up of Amundsen, Ellsworth, Nobile, the navigator, Captain Hjalmar Riiser-Larsen, and seven other Scandinavians, including the Swedish meteorologist, Finn Malmgren, and the five Italian mechanics. The pole was reached without incident and the flags of Norway, the United States and Italy dropped on the ice. During the passage over the unknown tract between the pole and Alaska, visibility was impaired by fog, but Amundsen was able to determine that no large land mass occurred in that area. Ice fragments flying from the propellers damaged the outer envelope of the airship and caused some anxiety for the gas tanks within. This fear proved groundless and after forty-two hours of flight Riiser-Larsen's precise navigation brought the *Norge* to the Alaska shore barely ten miles from her target, Point Barrow.

Then troubles began, foreshadowing the later *Italia* disaster. The intention was to follow the southwesterly trend of the shoreline to Point Hope and then strike south for the landing place at Nome. But in gale and fog Riiser-Larsen lost his way and in fear of crashing into a mountain he directed the *Norge* westward until a gap in the fog showed them far out to sea with white-crested waves foaming beneath them. At Riiser-Larsen's request Nobile increased altitude in order to clear the fog and get a glimpse of the sun. In the solar heat the gas expanded until pressure gauges pointed to danger: Nobile's orders given in Italian were not promptly understood; some stir and confusion resulted before the nose of the dirigible would be depressed to bring her down to a safer altitude. It was at this point that Nobile is alleged to have lost his head. Eventually they recovered the Alaskan shore and began to grope southward for Nome, but, with all personnel spent with sleeplessness and the strain of the recent crisis, it was determined to land at the first convenient spot. Despite the gale Nobile brought the airship to a safe landing at the village of Teller. The crew were carried on to Nome, a hundred miles to the south, and from there taken by steamer to Seattle.

The subsequent differences arising between Amundsen and Nobile will be noticed here only as far as justice to the unfortunate and much maligned Italian requires. However lacking in courtesy, Amundsen was within his abstract rights in excluding the foreign general from a share in the titular leadership of the expedition: he had justification and good business reasons for trying to bar him from the Press and the lecture-hall. But Nobile cannot be blamed for rebelling against the subordinate role imposed on him and for using the concessions obtained from Amundsen's agents to write articles for the newspapers and deliver lectures on the flight of the *Norge*. Amundsen retaliated by a newspaper article, later expanded in his autobiography, in which he coupled one or two serious charges against Nobile with a host of complaints so malicious and petty that the judicious reader is more offended at the accuser than at the accused. He also pressured Riiser-Larsen into endorsing his ignoble denigration of a comrade.

Nobile did not learn the full text of this indictment until after he had crashed in the *Italia* and Amundsen had died in attempting his rescue. In decency he held his tongue for many years until, learning that the slanders against him were still current, he published a temperate and generally convincing apologia. Though "clearly no man enjoys being held up to ridicule as a compound of vanity, egotism and self-seeking," he disregards all personalities and confines himself to proving his technical competence, which he does in a satisfactory manner, if allowing for the effects of prolonged sleeplessness and exhaustion. He quotes Amundsen's own testimony to the skill with which "having slept only 3 hours out of the preceding 97" he brought the Norge to a soft landing at Teller.

The *Norge*'s flight seemed to prove the fitness of the dirigible for surveying the large extent of the north polar region still unvisited by man. Amundsen, now in his fifty-eighth year, declared his career as an explorer ended. But Nobile's ambition was aroused; while still at Teller he invited Riiser-Larsen to share in the leadership in another voyage devoted to systematic exploration of the northern ocean. This promising partnership was nipped in the bud by the quarrel with Amundsen, and Nobile was left to depend on the resources which his own country could provide. He obtained the sponsorship of the Royal Italian Geographical Society, secured the use of the airship *Italia*, sister to the *Norge*, for the flight and that of the naval training vessel, *Citta di Milano* for the conveyance of materials and groundcrew to his Spitsbergen base.

Nobile's project did not receive the welcome in the press of Europe and America that its ambitious but practical content deserved. In substance his plan was the logical sequel to the experimental flight of the *Norge*, and promised richer and more varied gains in the sphere of scientific knowledge. Amundsen had carried only one professed scientist, Malmgren, on the *Norge*; Nobile embarked three on the *Italia*, Malmgren, the Italian, Aldo Pontremoli, and the Czech, Francis Behounek, to supplement the geographical survey with research in oceanography, terrestrial magnetism and radioactivity. But the enterprise originated in Fascist Italy; the dislike, founded on fear and mistrust, which foreign journalists felt for the Head of State, Mussolini, was extended to Nobile, whom they supposed erroneously to be one of the Duce's favorites. Amundsen's enormous prestige made his strictures on the Italian damaging, and gave rise to the idea that the expedition was a vain and presumptuous attempt to recover a supposedly forfeited reputation. The proposed ceremony at the North Pole, purely incidental to a landing there for scientific observation, was given a prominence calculated to excite ridicule. The general tone on news items was cold rather than hostile, but it was evident that Nobile would be harshly used by the organs of public opinion unless this journey was one of unadulterated success.

It very nearly was. At 1 P.M. on 15 May 1928, the *Italia* took off from King's Bay to Franz Josef Land and, rounding Cape Fligely, continued on an easterly course to fix the still unexplored western shores of the Severnaya Zemlya group. Hindered by gales and concerned for his petrol supply Nobile stopped a little short of this objective, but enlarged his discoveries by sweeping down southwest to Cape Zhelaniya, the northern point of Novaya Zemlya, before returning to his base. In a flight of sixty-nine hours the *Italia* had flown 2,400 miles, and lifted a tract of 17,250 square miles of ocean from its primeval obscurity.

On May 23 the airship was off again in the opposite direction. In pursuance of the methodical program he had drawn up Nobile flew to Cape Bridgman, the northeast angle of Greenland, and turned north, thus running a skewer through the heart of the unexplored sector lying between Peary's 1909 sledge track from Cape Columbia to the pole and Byrd's flight route in 1926 from Spitsbergen. With a strong following wind the *Italia* covered the 425 miles from Greenland to the top of the globe at sixty miles an hour, arriving twenty minutes after midnight, May 24. As the gale blew too strong for the scientists to make their intended landing for observations on the ice, the dirigible circled in the air while the Italian flag, the gonfalon of the city of Milan, a religious medal and a wooden cross, blessed by Pope Pius XI, were dropped on the ice.

The wind continued high and almost dead foul for a flight back to Spitsbergen. Nobile conferred earnestly with Malmgren on the best course to pursue in these circumstances. He himself favored sailing on with the wind to Siberia or alternatively making for the Canadian shore where the meteorological broadcasts promised a calm atmosphere. Malmgren strenuously opposed both these options. Failure to return to their base would entirely upset the scientific program; and he was confident that the gale from the southwest would soon drop and be replaced by winds from the northwest. Desirous on his own account of the exact fulfillment of his mission, the commander smothered his scruples and gave the order to head back for King's Bay. The decision was Nobile's, and with a generosity and delicacy which his own critics would have done well to imitate, he absolves Malmgren from blame for the ensuing catastrophe. The wind did change as the meteorologist forecast, but not as soon as he expected; "but who can tell whether" on the flight to the Mackenzie River "some other perils would not have been lying in wait for us?"

Leaving the pole at 2:20 A.M., May 24, Nobile steered a course several degrees east of the direct line to his base to avoid head-on encounter with the wind and to obtain original observations to the east of Byrd's flight-track. For some time the *Italia* sailed blind in cloud and fog, steering by magnetic compass. As speed could not be measured by the rate of the motors in a storm, she dived below the fog to an altitude of 600 feet: it was ascertained that the wind quartering on her nose was cutting her speed down to twenty-six mph and giving her a pronounced sidedrift. She sailed on through fog patch and snow squall, gathering ice on her outer covering as the day advanced. Concerned at their slow progress against a wind varying from twenty to thirty mph, Malmgren urged Nobile to put on speed and get as quickly as possible out of the storm zone. Nobile raised the airspeed to sixty mph netting a groundspeed of thirty-seven in the teeth of a high wind, punctuated by sudden squalls which drove the aircraft up to thirty degrees off course. Alarmed at the rapid consumption of petrol and the strain which high airspeed was putting on the light but bulky structure of the airship, he slowed his motors, but on Malmgren's remonstrating again accelerated them.

The *Italia's* blundering and zigzag course left the navigating officer in some doubt as to his whereabouts. Wireless signals exchanged with the *Citta di Milano* at King's Bay gave the direction to the base but left its distance a matter of uncertainty, though dead reckoning placed them close to the Spitsbergen north shore. To resolve this doubt they climbed above the fog at 3,300 feet and swept the sunlit wilderness of vapor to the southward with field glasses in vain. The high peaks of Spitsbergen were not yet in sight. The *Italia* dived back through the fog to a 900-foot level and cruised on, checking speed and drift by the pack below.

It was 10:30 in the morning of May 25 when Nobile heard Chief Technician Cecioni exclaim, "We are heavy." The instruments showed the airship down by the stern and falling rapidly. Nobile shouted to the rigger to check the stern gas valves, ordered the engines accelerated and the nose of the ship lifted—to no effect; she was still falling. Below them lay a ragged pack, its masses growing larger every moment of their slanting descent. At 10:33 the *Italia* crashed. A great part of the cabin was torn off, spilling ten of its occupants with equipment and rations on the ice. It is possible that some of the six men still aboard lingered deliberately to throw more supplies onto the ice. In the mist-shrouded wilderness of grey they did not perceive that the airship, lightened by the wreck of the cabin and also possibly by ice cover dislodged in the shock, was again floating up and off with the wind. Nobile lying stunned on the ice recovered consciousness to see his *Italia* drift off over the ice and fade from sight in a brooding fog bank.

Of the ten men thrown on the ice, one had been killed outright. Cecioni had broken his leg and Nobile was badly crippled by injuries. Malmgren's right arm and side were badly bruised. The other six, the naval officers, Mariano, Zappi and Viglieri, the engineer Trojani, Dr. Behounek, and the wireless operator Biagi had come off with none or only minor injuries. A search of the wreckage strewn on the ice yielded a small tent, sleeping bags and enough food to support the party for a month, while the recovery undamaged of astronomical kit and field wireless set offered good prospect of rescue within that period. Observations thus made possible placed the crash at 81° 14' latitude north, and 28° 14' east. They were some sixty miles from the north shore of Northeast Island, Spitsbergen, and quite close to off-lying islets.

Biagi quickly improvised a wireless mast and began transmitting the SOS signal every other hour, for several days without success, owing to some defect in his apparatus or to inattention on the part of the *Citta di Milano*. After some hours when the *Italia* was plainly overdue, her captain began to signal in his turn, that, supposing that the airship had come down near the north shore of Svalbard (Spitsbergen), he was organizing a search between the 15° and 20° meridians of east latitude. But Nobile had deliberately chosen a course east of the King's Bay meridian, sidedrift had carried him even further to the left: he and his comrades were lying on the pack many miles east of the designated search area, and, to make matters worse, they quickly perceived that wind and drift were carrying them even further off to the southeast toward the waters where Andrée (of whose fate they were still unaware) had perished. Their chances of rescue, already slender if the wireless refused to work, seemed to be diminishing daily.

The able-bodied members of the party chafed at inaction. Two days after the crash Mariano and Zappi proposed that they along with Malmgren (injured, but the only man with experience in travel over the ice) should attempt to make their way by foot to the shore and on to King's Bay. Nobile opposed the plan, urging that the journey over broken and melting ice was impossible for men without sledges who must carry the means of shelter and subsistence strapped on their backs, and that to wait for the wireless to become effective, though a poor resource, was the only one left them. The two officers were unconvinced—Biagi, arguing that the wireless was plainly useless, offered to join them, provoking others to join in the scramble for supposed safety. Behounek stopped this unseemly auction by declaring that on no account would he desert the General; and Malmgren (who was regarded as indispensable to the traveling party) refused to budge if Nobile and Cecioni were to be left with insufficient care and protection. This implied rebuke was enough to recall the good Biagi to a sense of duty. Mariano, Zappi and Malmgren only, with Nobile's reluctant consent, took their departure carrying on their backs their share of the food store. Though his burden was the lightest of the three Malmgren probably foresaw that the bruised side which had permitted him to do routine work about the camp would bring on crippling exhaustion on a long and sustained march. On taking leave of Nobile he gave it as his secret opinion that "both parties would perish."

Two developments in the next few days gave the able-bodied men in the camp reason to be thankful that they had not shared in the journey. The first was the discovery that the movement of the ice which had so alarmed them was not constant but largely dependent on the wind. And then on June 6 Biagi intercepted a signal which proved that his distress signal had got through at last. On June 3, nine days after the crash, it had been picked up by a Russian amateur at Archangel in a garbled but recognizable form, and transmitted via Moscow to Rome. Communication was established with the *Citta di Milano*—until now a heedless listener because Captain Romagna supposed the wrecked party unable to transmit; Nobile learned that the aircraft and pilots (including Riiser-Larsen) of several nations and the Soviet icebreaker *Krassin* were en route to the rescue; in turn he transmitted the bearings of his camp and detailed instructions for locating it. This proved no easy matter on the broken and discolored surface of the melting pack. Four times the castaways had the mortification of seeing rescue planes fly close overhead without observing them. On June 20 an Italian flying boat after much dubious circling spotted them and dropped supplies. On the twenty-third the Swede Lundborg after painstaking reconnaissance was able to land his ski plane not far from the camp.

Anticipating a piecemeal rescue Nobile had prepared a roster in which he assigned himself fourth place, prior only to the navigator, Viglieri, and the wireless operator Biagi, who must of necessity remain to the last to ascertain and transmit the bearings of the camp if the ice happened to shift. But Lundborg on reaching the tent stated that his orders were to bring off the general and no other. Nobile replied that that was contrary to his duty as commander, and that the crippled Cecioni should be the first to go. Lundborg reiterated his orders, adding that Nobile was needed to furnish directions for finding the airship and the six missing men. (These instructions had been sent by wireless but not properly understood.) Cecioni, said the Swede, was too heavy a man to be taken aboard along with his observer. Reflecting that all would be rescued in a few hours, Nobile yielded. Bearing in mind his responsibility to the missing six, his compliance can hardly be blamed, though it was to cost him dear. Lundborg carried him to the Swedish base at Murchison Bay, but on his next (solo) flight to the tent, he capsized and wrecked his Fokker, fortunately without injury to himself, and joined the other captives in the tent. Further rescue flights were deferred until the lighter Moth plane could be procured. Nobile, now on board the *Citta di Milano*, found that unhappy circumstances had made him liable to the odious imputation of having deserted his men.

He also received the distressing intelligence of the disappearance of Amundsen over the Barents Sea. The cross-grained but very gallant explorer, neglecting in Nobile's urgent crisis the careful preparation which he had invariably practised, had come up from France with a crew of five in a flying boat of doubtful reliability. They refueled at Tromsø and took off again to be seen no more. Search for them, as for the six men on the *Italia*, proved fruitless.

The tent party were more fortunate, though their rescue was long deferred. A Swedish pilot took Lundborg off in a Moth but found the ice so rotten that he dared not repeat the flight. The eddying pack brought the stranded men to within four miles of Cape Leigh Smith and then swung them fifteen miles out to sea. In this impasse Nobile derived some satisfaction from the feeling that it was his presence on the *Citta di Milano* and the advice he gave that deterred the *Krassin*, which had damaged rudder and propeller in the ice, from leaving the work of rescue to the aircraft and withdrawing for repairs. After a period of detention in the pack near Seven Islands the Soviet icebreaker began to make headway on July 11, and grind her way east-ward,—the symbol in her sudden mastery of baffling circum-stances of coming Russian supremacy in the ocean where Anglo-Saxon and Scandinavian had until then held the stage. At 5:20 A.M. on the twelfth, a man was seen waving from the ice. It was Zappi; near him lay Mariano; both men were taken aboard in an advanced stage of starvation. Malmgren had died a month before. Late in the day the *Krassin* reached the tent camp and picked up the last five of the *Italia*'s survivors, forty-eight days after the crash.

The tale told by Mariano and Zappi furnished the world press with the luxury, until then lacking in the story of the *Italia*, the element of horror. From the very beginning of their journey they had been baffled by the movement of the ice. After weeks of exertion they were farther from their objective than when they left the camp. On the fourteenth day Malmgren declared himself unable to go further: he begged his comrades to dig a trench in the ice and leave him there to die. The Italians com-plied with his request, and walked on a hundred yards, and then, shame getting the better of the instinct for survival, they sat down and waited in the hope that Malmgren might recover his courage and rejoin them. After twenty-four hours the Swede raised his head from the trench and bade them, "Go on." This was the last they saw of him. As the two Italians journeyed on Mariano was stricken with snow blindness; Zappi soon wearied of helping him over crevasses and hummocks; in their turn they dug a shelter and waited for death. None of the planes flying overhead observed their signals. On July 4 the revolving pack brought them close to Broch Island; Mariano urged his com-panion to go on and save himself; Zappi had neither the will nor the strength to comply. They had lain twelve days without food when they were aroused by the siren of the approaching *Krassin*.

Amundsen's generosity (to his own disadvantage in some quarters) to the erring but cruelly used Frederick Cook encourages the belief that, had he lived, he would have used his enormous prestige to shield Nobile from the storm of obloquy which the termination of his enterprise provoked. An informed and benevolent advocate could have urged much on his behalf. He had embarked on a useful, honorable and distinctly hazardous mission, and after more than five days of actual flight had missed by only hours a complete and flawless success. But the flight of the *Italia* marked the transition from glamorous and easily publicized discovery to the unspectacular accumulation of scientific data. Nobile's achievement in the latter area had not, like Payer's, been illuminated by the discovery of new land on which popular interest could focus. The only newsworthy feature of his voyage was the ceremony at the pole which was taken out of context and made to appear ridiculous. The superficial student of the news media in April and May 1928 was apt to suppose that the sole purpose of the flight was to drop religious and national symbols at the pole; and, if the author's recollections do not mislead him, one American scholar undertook gravely to refute the hypothesis that by this gesture the global cap was added to the domains of Fascist Italy. Only the catastrophe was available as a source of copy. The crash, seeming to justify the strictures of Amundsen, enabled the press to impugn Nobile's competence; his separation from his men—admittedly unfortunate—invited them to arraign his character. It was suggested with truly devilish malice that he had broken his leg not at the time of the accident but in sprinting over the ice to be the first to board the rescue plane of Lundborg. In similar vein the rumor was propagated that Mariano and Zappi had been guilty of cannibalism, a beastly slander, which the author could have no demonstrable reason for advancing and which the victims had no means whatever of disproving. The Italian government saved face by disowning the expedition and discountenancing its author. Nobile dared publish no apologia until after the collapse of the Fascist regime. His *My Polar Flights* appeared in English in 1961. As an undocumented piece of controversy, written by an old man a generation after the event, it is inconclusive, though it leaves a pleasant impression of a kindly, not too forceful man, who is careful to redeem his own character with the minimum of damage to the reputations of others. Nobile emerges with credit from a task to which neither training nor early choice had prepared him. He furnished the one indispensable factor in the Amundsen-Ellsworth flight, and was perhaps used with less than due consideration by those rugged, publicity conscious individualists. His own flight would have marked a significant stage in the development of aerial geography had not the dirigible been superseded through its own incurable defects and the evolution of the helicopter; his concrete achievement gives its author a place in the high European Arctic sequence of De Long, Nansen and Cagni. Misfortune, even without positive merit, entitles a man to our sympathetic regard. Umberto Nobile was both unfortunate and meritorious, and not unworthy of high rank in history as the last of the polar adventurers.

For Nobile was the last of the polar adventurers. His crash and Amundsen's death were a striking finale to the drama opened four centuries previously by the traders of London and Amsterdam. There was no more scope in the frozen seas for the amateur or semiprofessional. The drift of the *Fram*, the foot journeys of Cagni and Peary, and the flights of Amundsen and Nobile had revealed no new land, but a frozen waste, neither inviting nor very accessible to the private adventurer. The technical revolution has come to both polar zones: their secrets are being brought to light by groups of scientists in the employ of governments which alone can equip them, and by radio, aircraft and icebreaker, enable them to pursue their researches without prohibitive risk. Dangers and hardships are still plentiful, but they are far removed from the trials and hazards which give the toils of De Long, Nansen and scores of other polar travelers of the past so high a rank in the scale of human enterprise and fortitude.

BIBLIOGRAPHY

Abruzzi, Luigi, Duke of. *On the "Polar Star" in the Arctic Sea.* London: Hutchinson, 1903.

Adams-Ray, Edward, trans. *Story of Andrée.* New York: Blue Ribbon Books, 1931.

Amundsen, Roald, and Ellsworth, Lincoln. *First Crossing of the Polar Sea.* New York: George H. Doran, 1927.

Beaglehole, J. C., ed. *The Journals of Captain James Cook.* Vol. 3. Hakluyt Society. Cambridge University Press, 1967.

Berg, L. S. *Otkritiye Kamchatki i ekspeditsi Beringa, 1725–1742.* Moscow-Leningrad, 1946.

Coxe, William. *Account of Russian Discoveries between Asia and America, etc.* . . 4th ed. London, 1803.

Danenhower, J. W., *Narrative of the "Jeanette."* c. 1883.

De Long, Emma. *Voyage of the "Jeanette".* Boston: Houghton & Mifflin, 1884.

De Veer, Gerrit. *The Three Voyages of William Barents, 1594, 1595 and 1596.* Hakluyt Society. London.

Fiala, Anthony. *Fighting the Polar Ice.* New York: Doubleday, 1902.

Gilder, W. H. *Ice Pack and Tundra.* New York: C. Scribner & Sons, 1883.

Golder, F. A. *Russian Expansion on the Pacific, 1641–1850.* Cleveland: Arthur H. Clark, 1914.

Golder. *Bering's Voyages.* American Geographical Society, 1922.

Hakluyt's Voyages. Everyman.

Jackson, F. G. *A Thousand Days in the Arctic.* New York: Harper's, 1899.

Murphy, Robert. *The Haunted Journey.* New York: Doubleday & Co., 1961.

217

Melville, G. W. *In the Lena Delta.* Boston: Houghton Mifflin, 1884.

Nansen, F. *Farthest North.* London: Constable & Co., 1897.

Nobile, Umberto. *My Polar Flights.* Translated by Frances Fleetwood. London: Frederick Muller, 1961.

Nordenskiold, A. E. *The Voyage of the "Vega."* Translated by A. Leslie. London: Macmillan, 1883.

Parry, Ann. *Parry of the Arctic.* Fernhill, 1963.

Payer, Julius. *New Lands Within the Arctic Circle: A Narrative of the Discoveries of the Austrian Ship "Tegethoff" in the Years 1872–1874.* London: Macmillan, 1876.

Viese, V. Yu. *Morya Sovetskoi Arktiki: Ocherki po istorii issledovania.* Moscow-Leningrad: 1940.

Waxell, Swen. *The American Expedition.* Translated by M. A. Michael. London: Constable, 1950.

Wrangell, Ferdinand von. *Narrative of an Expedition to the Polar Sea in the Years 1820–23.* Translated by E. Sabine. London: 1840.

INDEX

Abruzzi, Luigi Amedeo, Duke of, heads Italian *Stella Polare* expedition to Franz Josef Land, 186, fixes winter quarters at Teplitz Bay, 186, disabled by frostbite, 187, assigns command of polar journey to Cagni, 187.

Adak, island in Aleutians, Chirikov at, 71.

Agattu, Aleutian Island, sighted by Chirikov, 71.

Alaska, discovered by Gvozdev, 55, coastal survey by Cook, 89–90, objective of Amundsen's transpolar flight in Norge, 204–5.

Ah Sam, cook, on Jeannette, 127, dies with De Long, 143–4.

Aldan, East Siberian river, 38, 53.

Alexandra Land, in west of Franz Josef Land, 168, 192.

Alexeev, Fedot, with Dezhnev on Kolyma, 41, lost at sea, 41.

Alfred the Great, saxon king of England, records voyage of Ohthere to White Sea, 4.

Alger Id., Franz Josef group, site of Camp Ziegler, 194.

Ambler, Dr. J. M., surgeon of *Jeannette,* 115, services on trek over the ice, 120–5, death in Lena Delta, 143–4.

America, under Baldwin, 194, and Fiala in Franz Josef Land, 194, wrecked, 195.

Amundsen, R., plans transpolar flight by airship, 202, reconnaissance flight by plane, 202, crosses N. pole in *Norge,* 203–5, differences with Nobile, 205, death, 212.

Anabar, North Siberian river, 60, 152.

Anadyr, river of NE Siberia, Dezhnev on, 40–43, 65.

Andrée, S. A., Swedish scientist and aeronaut, projects balloon flight over pole, 177, disappears on flight, 180, remains found, 180–84.

Anjou, with Wrangell on Siberian north shore, 93, explores New Siberian Ids., 93, 96.

Ankudinov, Cossack with Dezhnev on voyage E. from Kolyma, 41.

Armitage, Lieut., A., with Jackson in Franz Josef Ids., 170, on sledge journeys, 170, 171, 175, on boat voyage, 170–1.

Attu, Aleutian island, Chirikov at during search for Bering, 73.

Auk, Cape on Prince Rudolf Id., 107, 193.

Austria Sound, Channel in Franz Josef group, traversed by Payer, 106–9.

AVACHA, bay on SE coast of Kamchatka where harbour of Petropavlosk established, 67.

Baldwin, in Franz Josef Ld., 194.

Barents, W., Dutch navigator, takes ship to explore Novaya Zemlya, 1594, 21, Chief Pilot on large expedition to Vaygach Strait, 1595, 21–2, Chief Pilot to Heemskirk, 1596–7, rounds N. end of Novaya Zemlya, 23, winters at Ice Harbour, 24–8, dies on boat voyage homeward, tribute to, 33.

Barkin, Cape, on Lena Delta, 127, 130.

Bartlett, Capt. R., 90.

Bartlett, stoker on *Jeannette,* advises Mel-

219

Index

Index

Discovery in Russian and Siberian Waters

Index